DIGITAL RUBBISH

a natural history of electronics

Jennifer Gabrys

The University of Michigan Press ❧ *The University of Michigan Library*
Ann Arbor

Published in the United States of America by
The University of Michigan Press and
The University of Michigan Library
Manufactured in the United States of America
∞ Printed on acid-free paper

2014 2013 2012 2011 4 3 2 1

A CIP catalog record for this book is available from the British Library.

Library of Congress Cataloging-in-Publication Data

Gabrys, Jennifer.
Digital rubbish : a natural history of electronics / Jennifer Gabrys.
p. cm.
Includes bibliographical references and index.
ISBN 978-0-472-11761-1 (cloth : alk. paper)
1. Electronic waste. 2. Electronic apparatus and appliances—History. I. Title.
TD799.85.G33 2011
363.72′88—dc22 2010033747

end of life. Scan any city street, and you may find discarded monitors and mobile phones, printers and central processing units, scattered on curbsides and stacked in the dark spaces between buildings.

These remainders accumulate into a sort of sedimentary record, from which we can potentially piece together the evolution and extinction of past technologies. These fossils are then partial evidence of the materiality of electronics—a materiality that is often only apparent once electronics become waste. In fact, electronics involve an elaborate process of waste making, from the mining of metals and minerals, to the production of microchips through toxic solvents, to the eventual recycling or disposal of equipment. These processes of pollution, remainder, and decay reveal other orders of materiality that have yet to enter the sense of the digital. Here are spaces and processes that exceed the limited transfer of information through hardware and software. Yet these spaces and processes are often lost somewhere between the apparent "virtuality" of information, the increasingly miniature scale of electronics, and the remoteness of electronic manufacture and disposal. It is possible to begin to describe these overlooked infrastructures, however, by developing a study of electronics that proceeds not from the perspective of all that is new but, rather, from the perspective of all that is discarded.

Where does all the electronic detritus go once it has expired? The theory of waste developed in this book describes processes by which electronics end up in the dump, as well as what happens to electronic remainders in their complex circuits prior to the dump. Just as there are material, social, and economic infrastructures that support the growth and circulation of electronics, so, too, are there elaborate infrastructures for removing electronic waste. Underground, global, and peripheral residue turns up in spaces throughout the life and death of electronics. This study considers how electronics migrate and mutate across a number of sites, not only from manufacture to disposal, but also across cultural sites spanning from novelty to decay. My intention is to crack open the black box of electronics[1] and to examine more closely what sediments accumulate in the making and breaking of electronics. Yet, by focusing on waste, this book is less interested in material comprehensiveness, or all that goes into electronics, and is instead more attentive to the proliferations—material, cultural, economic, and otherwise—that characterize electronics. There is much more to electronics than raw materials transformed into neat gadgets that swiftly become obsolete. Electronics are bound up with elaborate mechanisms of fascination, with driving economic forces beyond the control of any single person, and with redoubling rates of innovation and decay.

In a time when media occupy our attention most unmistakably when they are present as *new* media, a study of dead media would, presumably, begin to describe the invisible resources expended and accumulated in these interlocked ecologies. In his "dead media" proposal, Sterling calls for a paleontological perspective, an approach that would account for the extinctions and sedimentations of lost media technologies, perhaps even with the object of preventing past media mistakes. To pursue this project, I have opted to develop a more particular natural history, which examines outmoded electronics as "fossils" that bear the traces of material, cultural, and economic events. Rather than amass a collection of outdated artifacts, then, this natural history suggests it is necessary not to focus solely on abandoned electronic gadgets but also to consider the extended sites through which electronics and electronic waste circulate, as well as the resources that assemble to facilitate these circulations. This natural history works not, however, from the assumption of never-ending technological evolution and progress but, rather, from the perspective of transience. What do continual cycles of novelty and obsolescence tell us about our material cultures, economies, and imaginaries? What other stories might emerge from the fossils of these obsolete commodities? In the end, this is not the handbook that Sterling describes. It is not an encyclopedic item that features so many odd but strangely attractive dead media. Instead, with any luck, it is the sort of study that, through another natural history method, traces the fossils of digital media within more heterogeneous material, political, and imaginary registers, while also providing insights into the complex ways that electronics fall apart.

The topic of electronic waste is situated at the intersection of a number of disciplines and locations. While this project dates to doctoral research begun in 2002, it also has a longer span of interest from the time I spent practicing landscape architecture and conducting fieldwork, design, and research in waste sites in North America. During the course of researching, writing, and revising this text, numerous people, from electronics recyclers to archivists of computing history, have extended support to the project. While not an exhaustive list, I would like to thank faculty and graduate students (past and present) in the Department of Art History and Communication Studies at McGill University, including Will Straw, Sheryl Hamilton, Darin Barney, Cornelius Borck, Jonathan Sterne, Christine Ross, and Jasmine Rault. Faculty members at Concordia University in Montreal have also provided valuable help along the way, including Johanne Sloan, Kim Sawchuk, Michael Longford, and Lorraine Oades.

This project has been made possible and greatly enhanced by funding received from several sources, including the McGill Majors Dissertation Fellowship; the Mellon Foundation Dissertation Fellowship through the Institute of Historical Research in the School of Advanced Study at the University of London; the Social Sciences and Humanities Research Grant for dissertation fieldwork and research through the Research Grants Office at McGill University; the Researcher in Residence program at the Daniel Langlois Foundation, Centre for Research and Documentation (CR+D) in Montreal; a dissertation fellowship from the Center for Research on Intermediality in Montreal; and the Design Department at Goldsmiths, University of London, which provided research and publication assistance. While a researcher in residence at the Langlois Foundation, I developed a wider view of electronic culture and art through reviewing the holdings at the CR+D. I would like to extend my appreciation to everyone at the center, including Vincent Bonin, Alain Depocas, and Jean Gagnon.

With funding from the CR+D, I was further able to visit numerous recyclers of electronic waste in the United States and Canada. I would like to thank individuals from Envirocycle, Back Thru the Future, Waste Management and Recycling Products, Retroworks, and Per Scholas for providing me with tours of their facilities and for explaining more about the complexities of electronics recycling. The recycling practices described in this study are informed by, but do not necessarily directly describe, the operations of these individual businesses. Other individuals who have helped in the research and fieldwork for this project include Megan Shaw Prelinger at the Prelinger Archives in San Francisco, Penny McDaniel at the U.S. Environmental Protection Agency in San Francisco, Bette Fishbein at INFORM in New York City, and Francis Yusoff in Singapore. Thanks are also due to everyone involved with the "Zero Dollar Laptop" project in London, including Ruth Catlow of Furtherfield, Jake Harries of Access Space, and participants from St. Mungo's charity for the homeless, for having me as a guest during their project press launch.

The funding received in support of this research allowed me to visit archives of computing history and to conduct fieldwork on electronics and electronic waste. I would like to thank archivists for their assistance in accessing holdings in computing history at the Smithsonian Institute in Washington, DC; the Computer History Museum in Mountain View,

California; the London Science Museum Computing Archives; the British Film Institute in London; the National Archive for the History of Computing in Manchester; and the Charles Babbage Institute at the University of Minnesota in Minneapolis. Tilly Blyth was especially helpful in facilitating my access to the holdings in computing history at the London Science Museum, and Stephanie Crowe made available a wealth of materials at the Charles Babbage Institute. Simon Lavington also provided a useful framework for understanding the history of computing while I was working in archives in the United Kingdom. While I was conducting archival research in London, Scott Lash at the Centre for Cultural Studies at Goldsmiths, University of London, graciously served as my mentor. Thanks are also due to faculty and graduate students at the Centre for Cultural Studies for the seminars and events that provided me with a collegial environment while researching in London.

I have received many helpful suggestions from colleagues at conferences and seminars where I have presented parts of this material, including the "Making Use of Culture" conference at the Cultural Theory Institute, University of Manchester; the "Ethics and Politics of Virtuality and Indexicality" conference at the Centre for Cultural Analysis, Theory, and History at the University of Leeds; the "Modernity and Waste" conference at the University of St. Andrews; and the "Design and Social Science" seminar series at the Centre for the Study of Invention and Social Process at Goldsmiths, University of London.

Portions of the introduction were published previously by MIT Press and *Alphabet City Magazine* as "Media in the Dump," *Trash* 11 (2006): 156–65; portions of chapter 1 by the MIT Department of Architecture as "The Quick and the Dirty: Ephemeral Systems in Silicon Valley," *Ephemera* 31 (2006): 26–31; portions of chapter 3 as "Appliance Theory," *Cabinet* 21 (2006): 82–86. Thank you to these presses and publications for permission to republish this material.

Finally, I would like to thank the anonymous referees for providing useful suggestions for revisions and the staff at the University of Michigan Press, including Tom Dwyer, Alexa Ducsay, and Christina Milton, for their guidance in all aspects of bringing this project to publication. I would also like to thank David Gabrys and Kathryn Yusoff, who have gracefully endured more than a few extended conversations and readings in relation to this text.

New shipment of electronic waste, Guangdong, China, 2007. (Photograph courtesy of Greenpeace / Natalie Behring-Chisholm.)

Contents

Introduction
A NATURAL HISTORY OF ELECTRONICS 1

1. Silicon Elephants
THE TRANSFORMATIVE MATERIALITY OF MICROCHIPS 20

2. Ephemeral Screens
EXCHANGE AT THE INTERFACE 45

3. Shipping and Receiving
CIRCUITS OF DISPOSAL AND THE "SOCIAL DEATH" OF ELECTRONICS 74

4. Museum of Failure
THE MUTABILITY OF ELECTRONIC MEMORY 101

5. Media in the Dump
SALVAGE STORIES AND SPACES OF REMAINDER 127

Conclusion
DIGITAL RUBBISH THEORY 147

Notes 159

Bibliography 201

Index 221

Silicon Valley Boulevard, 2005. (Photograph by author.)

Introduction

A NATURAL HISTORY OF ELECTRONICS

To each truly new configuration of nature—and, at bottom, technology is just such a configuration—there correspond new "images."

—WALTER BENJAMIN, "Convolute K," in *The Arcades Project*

The domain of machine and non-machine non-humans (the unhuman in my terminology) joins people in the building of the artifactual collective called nature. None of these actants can be considered as simply resource, ground, matrix, object, material, instrument, frozen labor; they are all more unsettling than that.

—DONNA HARAWAY, "The Promises of Monsters"

Electronic Waste

If you dig down beneath the thin surface crust of Silicon Valley, you will find deep strata of earth and water percolating with errant chemicals. Xylene, trichloroethylene, Freon 113, and sulfuric acid saturate these subterranean landscapes undergirding Silicon Valley. Since the 1980s, 29 of these sites have registered sufficient levels of contamination to be marked by the U.S. Environmental Protection Agency (EPA) as Superfund priority locations, placing them among the worst hazardous waste sites in the country.[1] In fact, Silicon Valley has the highest concentration of Superfund sites in the United States. What is perhaps so unexpected about these sites is that the pollution is not a product of heavy industry but, rather, stems from the manufacture of those seemingly immaterial information technologies. Of the 29 Superfund sites, 20 are related to the microchip industry.[2] The manufacture of components for such technologies as computers, mobile devices, microwaves, and digital cameras has contributed to the accumulation of chemicals underground. Mutating and migrating in the air and earth, these caustic and toxic compounds will linger for decades to come.

Silicon Valley is a landscape that registers the terminal, but not yet terminated, life of digital technologies—a space where the leftover residue of electronics manufacturing accumulates. Yet this waste is not exclusive to the production of electronics. Electronic waste moves and settles in circuits that span from manufacturing sites to recycling villages, landfills, and markets. Electronics often appear only as "media," or as interfaces, apparently lacking in material substance. Yet digital media materialize in distinctive ways—not just as raw matter, but also as performances of abundance—often because they are so seemingly immaterial. The elaborate infrastructures required for the manufacture and disposal of electronics can be easily overlooked, yet these spaces reveal the unexpected debris that is a by-product of the digital. The waste from digital devices effectively reorders our understanding of these media and their ecologies.[3]

"Waste is now electronic," writes Gopal Krishna in describing the escalating number of obsolete electronic devices headed for the dump.[4] This is the other side to electronic waste—not a by-product of the manufacturing process, but the dead product headed for disposal. E-waste—trashed electronic hardware, from personal computers and monitors to mobile phones, DVD players, and television sets—is, like the electronics industry, growing at an explosive rate. Electronics consist of a broad range of devices now designed with increasingly shorter life spans, which means that every upgrade will produce its corresponding electronic debris. In the United States, it is expected that by 2010, 3 billion units of consumer electronics will have been scrapped at a rate of 400 million per year.[5] Many of these electronics have yet to enter the waste stream. Of the hundreds of millions of personal computers declared useless, at least 75 percent are stockpiled.[6] Computer owners store the outmoded model as though there might be some way to recuperate its vanishing value, but the PC is one item that does not acquire value over time. At some point, stockpiled computers and electronics enter the waste flow. Most of these consumer devices are landfilled (up to 91 percent in the United States),[7] while a small percentage are recycled or reused. Recycling, moreover, often involves the shipping of electronics for salvage to countries with cheap labor and lax environmental laws. The digital revolution, as it turns out, is littered with rubbish.

While much of the attention to electronic waste focuses on the recycling and disposal of computers, these devices comprise only a portion of the electronic waste stream. The pervasiveness of electronics—the insertion of microchips into such a wide range of systems and objects—means that the types of waste that emerge from electronics proliferate. Microchips—or "computers on a chip"—recast the extent of computing beyond the medium-sized memory machines that occupy our desktops to encompass miniature devices and distributed systems. Microchips can be found in computers and toys, microwave ovens and mobile phones, fly swatters and network architectures, all of which contribute to the stock of electronic waste.[8] While the use of these devices differs considerably, the material and technological resources that contribute to their "functionality" have a shared substrate in plastic and copper, solvents and silicon. Electronics typically are composed of more than 1,000 different materials, components that form part of a materials program that is far-reaching and spans from microchip to electronic systems.[9]

This book raises questions about how to investigate electronic waste as a specifically *electronic* form of waste. In what ways do electronics pollute, and what are the qualities and dispersions of this pollution? Electronic waste is more than just a jumble of products at end of life and encompasses new materialities and entire systems of waste making. Wastes related to electronics give rise to entirely new categories of waste classification and ways of regulating waste. While the electronics industries may not consume as many hazardous materials by volume as heavy industry, for instance, no comprehensive criteria account for the *degree* of toxicity of materials used in the manufacture of electronics.[10] But the proliferation of electronics occurs as much in the form of "hardware" as it does in programs or "software"—those seemingly more immaterial forms of digital technology, from information to networks, that still inevitably rely on material arrangements. Electronics are comprised of complex interlocking technologies, any part of which may become obsolete or fail and render the entire computing "system" inoperable.

Current reports and studies generated on electronic waste specifically contend with its increase and control, as well as the environmental dilemmas that emerge with the exportation of waste.[11] While these studies provide invaluable information about the volume, distribution, and policies surrounding electronic waste, my overriding intention is to situate electronic waste within a material and cultural discussion of electronic technologies. Waste is not just sheer matter, so, arguably, the methods for studying waste might also account for *more than empirical* processes of waste making. The sedimentary layers of waste consist not only of circuit boards and copper wires, material flows and global economies, but also of technological imaginings, progress narratives, and material temporalities. Waste and waste making include not just the actual garbage of discarded machines but also the remnant utopic discourses that describe the ascent of computing technologies—discourses that we still work with today.[12] Exhuming these layers and fragments from an already dense record requires expanded definitions of what constitutes electronic waste, as well as inventive methods for gathering together stories about that waste.

In this study, I take into account the range of delineations for what constitutes electronic waste, and I further expand the definition of electronic waste to an examination of these material and cultural processes that facilitate and contribute to technological transience. To bring these multiple layers of electronics into play, this investigation registers how and where electronics transform into waste. Through waste, we can register the effects of these devices—the "materiality effects" as well as "the unintended, 'after-the-fact' effects" or "perverse performativity."[13] Electronics continually perform in ways we have not fully anticipated. Electronic waste, chemical contamination, failure, breakdown, obsolescence, and information overload are conditions that emerge as wayward effects of electronic materiality.[14] While these aftereffects are often overlooked, such perverse performativity can provide insights into technological operations that exceed the scope of assumed intentionality or the march of progress, and it can further allow the strangely materialized, generative, or even unpredictable qualities of technologies to surface.[15] Rather than move quickly to proposals for remedying these electronic dilemmas, I look more closely at the mutable qualities of electronics and evaluate the multiple ways in which these technologies fail and stack up as toxic remainders.

The advantage of focusing on electronics through remainder is that not just the effects but also the material, cultural, and political resources that enable these technologies become more evident in the traces of these fossilized forms. Such an approach interferes with—while taking up—the specters of virtuality and dematerialization, which often ensure that the material "supports" of electronic technologies are less perceptible.[16] But materiality is more than a support, and as this study suggests, virtuality consists not just of the *appearance* of immateriality. Virtuality, I suggest, can even enable more extensive consumption and wasting. When electronic devices shrink to the scale of paper-thin and handheld devices, they appear to be lightweight and free of material resources. But this sense of immateriality also enables the proliferation of waste, from the processes of manufacture to the development of disposable and transient devices in excess. Here, I take as my point of departure this proliferation of possible types of electronic waste. These waste traces sediment into a natural history of electronics.[17]

Natural History: A Material Method

Imagine any typical electronic device broken into pieces, scattered into assorted component parts, and cast across disparate sites. Microchip and screen, plastic casing and packaging, electronic memory, peripherals and formless debris—all these sift out from the black box of electronics. Distinct fossils are generated and cast off throughout the life and death of electronics. These fossils bear the traces of electronic operations; they accumulate into a natural history record. But this natural history and these fossils are not remainders from past ice ages. Instead, they are the recently petrified forms from rapidly succeeding technological epochs. These fossils are more than inert objects to be decoded. They are indicative of places and "processes of materialization"[18] that have sedimented into and through these residual forms.

Bruce Sterling's proposal (quoted in the preface) to undertake a paleontological examination of dead media was, in fact, previously implemented in a much different way by the twentieth-century German cultural theorist Walter Benjamin, who developed a particular "natural history" method by reflecting on the fossilized commodities in the obsolete arcades of nineteenth-century Paris.[19] Strange, extravagant, yet mundane and ultimately broken-down objects assembled within his natural history, including "the briefcase with interior lighting, the meter-long pocket knife, or the patented umbrella handle with built-in watch and revolver."[20] For Benjamin, decaying objects and outmoded objects that were no longer fashionable revealed concrete facts about past cultural imaginings. By examining these objects, it might be possible to discern not just their former lives but also the larger contexts in which they circulated, as well as the economic and material forces that contributed to their sedimentation and decay. His natural history presents a method for exploring the transitory impulses that unfold through commodities and technologies.[21]

Such a natural history is an effective guide for thinking through the remainders of electronic waste. But this is not a conventional rendering of natural history. The emergence of natural history as a more usual practice of classification and description signals, in Michel Foucault's account, the beginning of the "modern episteme."[22] From the seventeenth century onward, natural history increasingly operated as a process of "purification," where the allegorical dimensions of naming things and of forming stories about the natural world were erased from scientific practice. In this way, it became possible to represent an animal or vegetable objectively—without the intervention of myth or fable. Such transparent descriptions depended on established and often physical criteria (e.g., color or size) by which specimens could be identified. This practice of natural history has enabled a whole set of modern scientific practices that filter out the noise between words and things and that delete the "play" of calling the world into being through language.[23] Charles Darwin's particular development of a theory of evolution is situated within this longer natural history, but his observations have often been conflated with (Victorian) notions of progress[24]—the same notions of progress within natural history that Benjamin sought to challenge in his own natural history method.

Benjamin, in his practice of natural history, at once drew on but departed from the usual, more scientific practice of natural history. While he was fascinated by nineteenth-century depictions of and obsessions with natural history and fossil hunting, he interpreted these historical records of the earth's deep time as a renewed temporal vantage point from which to assess practices of consumption. Obsolete objects returned to a kind of prehistory when they fell out of circulation, at which time they could be examined as resonant material residues—fossils—of economic practices. He reflected on the progress narratives that were woven through Victorian natural histories (and economies) and effectively inverted these progress narratives in order to demonstrate the contingency and transience of commodity worlds.

In this natural history of electronics, I take up the suggestive and unconventional natural history method developed by Benjamin and extend it—laterally—not as a model to replicate and follow but as a provocation for how to think through the material leftovers of electronics. The natural history method allows for an inquiry into electronics that does not focus on either technological progression or great inventors but, rather, considers the ways in which electronic technologies fail and decay.[25] These failures and sedimentations can be understood in part through the repetitive urge to pursue technological progress and regularly "upgrade." By focusing on the outmoded, it is further possible to resuscitate the political and imaginary registers that are so often forgotten in histories that rely on the persistent theme of progress.

Outmoded commodities are fossilized forms that may—through their inert persistence—ultimately unsettle notions of progress and thereby force a reevaluation of the material present.[26] While commodities might guide us to a space of speculative promise, the vestiges of these promises are all around us. These fossils persist in the present even as the assumed progress of history renders them obsolete. Within and through these forms, more complex narratives accumulate, which describe technologies not only as they promise to be but also as they materialize, function, and fall apart. In this Benjamin-inspired natural history method, such an approach to fossilized commodities becomes a way to circumvent "naturalized" histories, which typically assume that technological progress is automatic and inexorable or even a "natural" event, on par with evolution. Histories of technological forms are often narrated through the logic of "onward and upward," of crude early devices eventually surpassed by more sophisticated solutions. But rather than examine technology as an inevitable tale of evolution, I take up the notion that these fossil forms are instead evidence of more complex and contingent material events.

This natural history method, then, signals a distinct approach to materiality—not just as raw stuff, but, rather, as materiality effects.[27] Electronic fossils are in many ways indicative of the economies and ecologies of transience that course through these technologies. Electronics are not only "matter," unfolding through minerals, chemicals, bodies, soil, water, environments, and temporalities. They also provide traces of the economic, cultural, and political contexts in which they circulate. To begin to develop a more material account of these dematerialized technologies requires accounting for the multiple registers of what constitutes materiality—not as the raw matter of unproductive nature made productive, nor even as "second nature,"[28] but as a complex set of material processes and relations.

What would it then mean to do a natural history of electronics, if the sense of natural history encompassed these complex conjugations of materiality, nature, and history and also accounted for the telling of histories not as progress narratives but as more embedded, deeply material, spatial, temporal, and political effects? In this way, the microchip, as one of the fossilized forms discussed here, can be conceived of as a site where materials, environments, bodies, politics, technologies, ecologies, and economies accumulate. The microchip appears to be a thing in itself, similar to the way in which Haraway describes the gene. This is the way in which commodities are fetishized; they seem to be free-floating and without consequence. Yet the microchip, like the gene, requires "all the natural-social articulations and agentic relationships," from "researchers" to "machines" and "financial instruments," in order to circulate in the world.[29] Discussing these "things" involves being able to register the complex forces that bring them "into material-semiotic being."[30] This study does not advocate an approach that attempts to de-fetishize the chip or electronics. Instead I seek to develop a method that can encompass the apparent singularity of the chip together with the things it powers and the disparate fields it affects.

In this material method, I attempt to develop a practice of thought that works through cast-off objects in order to take up the density and "scatter" of electronic materialities.[31] This is a method that, following Benjamin, focuses on the "micrological and fragmentary," in order to "relate them directly, in their isolated singularity, to material tendencies and social struggles."[32] Such a method of natural history is not prescriptive but, rather, works across fragments and fossils to material processes and social conditions. By encountering fragments as traces of material processes, it is possible, as Benjamin notes, "to approach, in this way, 'what has been' . . . not historiographically, as heretofore, but politically, in political categories."[33] By not accepting naturalized histories, it is possible to engage with the political and situated character of materialities, progress narratives, and definitions of history and nature.

Taking up this more fragmentary approach, I work with the notion of the machine in pieces—of the fossilized forms of microchips, screens and plastic, memory and peripherals—in order to examine how these fossil forms are not just material remainders and effects but also indicative of the changing relations and definitions of technology, culture, nature, and history. "Nature," as Judith Butler notes, "has a history."[34] This natural history does not describe a commodity world operating alongside a more essential nature (where commodities, histories, and economies become naturalized); instead, it transforms nature and culture, staging their collision and revealing their shared conditions of transience.[35] Shifting definitions of "nature" can be identified through the different ways in which fossils have been interpreted throughout time. Fossils operate as indicators of changes in the "interrelated conception of nature, culture and history."[36] At one time, these encrusted forms might be read for proof of the Deluge; at another, they were evidence of the progress of life. From these readings, it is possible to develop an understanding of nature not as an essential or original reference point but as historical matter. Nature is no longer a stable ground against which it is possible to describe the progressions of culture. Benjamin put forward a neat summation of this approach in *The Arcades Project:* "No historical category without its natural substance, no natural category without its historical filtration."[37]

Why is it important—in a study of electronic waste—to think through the history of nature and the nature of history? Distributions and definitions of nature are never static, and through their shifting registers and relations to "culture" and "history," these definitions also inevitably inform the politics of matter and processes of materialization. Nature, while historical, cannot be reduced to either sheer process of social construction or inert matter. Because it is historical, it is emergent, contingent, embodied, and political. It is not absolute, which is important to articulate when anything cultural comes to seem to be an absolute condition. Technologies, economies, and commodities may appear to be natural or naturalized. But this is because they operate through a whole set of what Butler calls "sedimented effects."[38] Material appears to be given—as matter—because it has stabilized or sedimented, as Butler writes, "over time to produce the effect of boundary, fixity, and surface we call matter." This is the "process of materialization."[39] The fossils I investigate are not just congealed electronics but also a contaminated mixture of nature, history, and technology. Fossils effectively work to *denaturalize* technology and its effects. In this way, it is possible to engage with materiality not just as materialization but also as ultimately prone to instability and breakdown.

Fossils—the remainders and residues of technology and media—are, then, potent forms that bear the imprint of events (both actual and imagined); they are traces of prior lives, events, and ecologies. Residual matter and the unintended consequences of technology have emerged as a topic of interest within contemporary media studies, as well as studies of science and technology. In the edited collection *Residual Media,* media theorist Charles Acland suggests that residuals allow expanded ways of engaging with media beyond the obligatory narratives of media revolutions.[40] Similarly, in his media-archaeological investigations into the "deep time" of media, Siegfried Zielinski begins with the "rubbish heaps" of media, to suggest that bundled into media are more complex temporalities and imaginings that exceed the simple or assumed progression toward advanced devices.[41] By decoupling histories of media and technology from progress, it is possible to examine the more complex temporalities and materialities that accompany distinct media technologies. Such extended terrains further resonate with what media theorists, from Marshall McLuhan to Friedrich Kittler, have called the "media environment"[42]—understood as the material conditions and discursive "networks" that constitute media[43] or as the set of processes and effects that even suggest that "there are no media."[44] Rather than isolated media objects, there are institutions, practices, and devices that—assembled together—enable media operations.

The fossils studied here do not assemble into a network, however, nor are they "actors" in a planar field of influence.[45] Rather than circumscribing systems, these figures open into spaces of relation and resonance.[46] Fossils are not abstract distributions but, rather, temporal sedimentations and transformations; they are mutable and contingent forms. From this perspective, users—as well as electronics waste workers—are also part of the materiality effects of electronic technology. However, the focus in this book is less on how users engage with a vast array of computing devices—particularly since waste workers, among others, often play a much different role as "agents" in their engagement with electronic technologies.[47] The material culture of electronics discussed here is not centered on users as manipulators of media content but, instead, focuses on how materialized workers, technologists, and consumers all emerge in relation to processes of electronic obsolescence and decay.

Materiality is a topic and focus that is now pervasive across multiple disciplines, from media studies to geography and science and technology studies. Given its concern with drawing out the complex material processes of digital media, this study is primarily located within media studies, but it also draws on writings within cultural geography and science and technology studies to analyze these technologies.[48] What becomes evident in these writings is a shared interest in describing how matter matters, and in this way multiple terms emerge that are used both similarly and dissimilarly. Material may rematerialize or dematerialize, it may be performative or transformative, or it may circulate in or as a network, system, or circuit. While this study does deploy these terms, it calls out the ways in which many of these terms have specific histories within computing and information theory. The histories of these terms are material histories as much as intellectual histories, and where relevant I discuss the ways in which these often apparently abstract terms work in quite specific ways in the digital realm.

It may be tempting to chart a sort of life-cycle analysis of electronics in order to track the comprehensive movement from raw material to waste product.[49] But I intentionally do not seek to understand the circuits of electronic waste through a life-cycle analysis, which would run the risk of appearing to be a tidy analysis of inputs and outputs to the neglect of both the material and imaginative residues that accompany electronics. Instead, the circuits I pursue are spatial and material instantiations of how electronics generate waste, whether in the form of chemical contamination or information overload. But there is more to expiration than just the guilt of discards. As Benjamin demonstrates, outmoded commodities "release" the imaginary and wishful dimensions that made them so compelling when first distributed as novel objects. Natural history, as a study of expiration, also engages with this mythic aspect of innovation. Any investigation into electronics would be incomplete if it did not account for this more fantastic register of technologies, as well as the ways in which technology does not constitute an orderly narrative.[50]

Electronic waste is a topic that challenges the methods and delineations used to describe it. Benjamin's natural history method suggests ways to mobilize the possible play of relations within material culture, economies, consumers, dreams, and politics. This is a natural history method that is simultaneously political and poetic, concrete and literary. Data is never devoid of dreaming. What registers as empirical matter bears an inevitable relationship to theories that would identify and describe that matter. Deciding what counts as empirical matter is also a process of materialization.[51] As much as it draws attention to the complex material effects of electronics and electronic wastes, this natural history method is ultimately a strategy for *rematerializing* electronics.[52] Electronics can be rematerialized both in the way their pasts accumulate—as fragmentary and the outmoded—and in the way ecologies, politics, and imaginings emerge from the rubble. Natural history—as a theory, practice, and method—brings together questions of materialities, time, politics, environments, technology, commodities, and imaginings; it also reorients the relations between nature, history, culture, matter, and time. This is a method for collecting material residues and for reorienting the histories and temporalities that emerge with technologies. It moves across scales, from the fossilized fragment to the temporal landscape. It tells material histories not as fixed, abstract, or essential but as dynamic, concrete, and entangled.

This natural history is grounded in the time of electronics, situated within a historical framework that primarily coincides with the development of the microchip, although it also draws on the longer postwar history of computing and automation. The material, references, and sites assembled in the following chapters draw on diverse sources relevant to electronics and the material economies and ecologies of which they are a part. While this method is rooted in fieldwork and draws on theoretical literature in technology, media, and material studies, it also engages with primary sources, including archived objects and documents, Web pages and online interviews with electronic "pioneers," reports by governmental and nongovernmental organizations, annual reports, newspaper articles, and popular commentaries, which together capture not just the material textures of electronic waste but also the material textures of language relating to electronics.

I explore the material-semiotic aspects of electronics by writing alongside these texts, in a further attempt to work with—and even transform—the "technophilic" and "technophobic" approaches that can emerge, at turns, in relation to electronics.[53] This project is neither utopic nor dystopic in its discussion of electronics, but it does draw on both the hyperbolic promises and informational and material excesses through which electronics are described. My intention is to move beyond a utopic/dystopic "e-mail address," as Haraway suggests when describing her attempts to forge another position in relation to cultural salvation-or-catastrophe discourses.[54] Similarly seeking to find another route around the steady oscillations between positive and negative renderings of cultural history, Benjamin suggests, "Overcoming the concept of 'progress' and overcoming the concept of 'period of decline' are two sides of one and the same thing."[55] Benjamin then makes a "modest methodological proposal" to find a new "positive element," where failure is not just the flip side to progress but, rather, offers an opening or rupture into other material relations and imaginings.[56]

When Benjamin undertook his investigations into the natural history of commodities, he did so in urban landscapes that emerged through accreted registers of consumption. He focused on the "dying arcades" of Paris, where "the early industrial commodities have created an antediluvian landscape, an 'ur-landscape of consumption.'"[57] In the arcades, "a past become space,"[58] he was able to imagine how commodities and technologies transformed into residues that contained traces of the resources, labor, and imaginations that went into these transformations. Similarly, electronic waste calls attention to the spatial and material infrastructures that support the transformations of these technologies. In addition to the texts, documents, and objects already discussed, I here focus on a number of key sites in which the remains of electronics can be studied. Fieldwork conducted in the gathering of these spatial stories has ranged from Silicon Valley to Singapore and from the Bronx to London. Superfund sites and museums of the electronics industry, shipping yards and electronics recycling facilities, computing archives, and electronics superstores and repair shops inform the content, texture, and structure of this study, which takes up natural history as much as a method as a theoretical point of inquiry.

To chart the multilayered spatial and material infrastructure of electronic waste, I have organized the chapters in this book around five sites in which distinct electronic fossils can be located. I unearth these fossils found throughout the life and death of electronics, in order to register the diverse resources, materials, and imaginaries that undergird this technology.[59] These sites and fossils are microchips in Silicon Valley; screens used in market transactions of the National Association of Securities Dealers Automated Quotations system (NASDAQ); plastics—in the form of housing, packaging, and more—as they move through the spaces of shipping and receiving, consumption and disposability; memory devices stored and at work in the electronic archive; and all the peripherals and scrap, from printed circuit boards to copper wires, which can finally be found in the landfill and salvage sites. These fossils and spaces of remainder each embody specific processes of electronic materialities and electronic waste. These are not just "waste sites" but also temporal zones that register the speed and volume of production, consumption, and disposal of digital technologies.

The aging electronics that occupy dumpsters and landfills register not just as fossils from successive upgrades but also as objects that circulate through a number of spaces in the process of their making and unmaking. Circulation, as described throughout this study, is a method both for mapping electronic waste as it congeals in and moves through diverse spaces and, at the same time, for registering the often amorphous or mutable arrangements of electronics and electronic residues.[60] This research describes not a "society of flows" but, rather, sites of unexpected accumulation. I take up these scraps and fossils in the sites where they are found, in order to think through the disparate effects, sedimentations, and imaginaries that inform the making and breaking of electronics.

This book begins with the perception that digital technology is light, postindustrial, or dematerialized. Worldwide, discarded electronics account for an average 35 million tons of trash per year.[61] Such a mass of discards has been compared to an equivalent disposal of 1,000 elephants every hour.[62] A colossal parade of elephants—silicon elephants—marches to the dump and beyond; suddenly, the immaterial abundance of digital technology appears deeply material. A considerable amount of waste is also generated at the point of electronics manufacture. Chapter 1 traces these economies of abundance and focuses specifically on the waste that emerges in the interrelated production of microchips, information, and environments. Through a study of these material relations, it is possible to examine how "overload" is a condition that describes information and contaminated environments alike.

Before it becomes trash, however, digital technology drives another type of abundance, this time in the dematerialized space of electronic trading. NASDAQ is the electronic trading market that specializes in technology companies, and it is also the world's first electronic stock market. Established in 1971, NASDAQ was described in its 2004 annual report "Built for Business" as the world's largest "electronic screen-based equity securities market." NASDAQ is an index of the volume and value of technologies, but it is also a digital technology of its own. As an automated system programmed to deliver financial data across a scattering of sites, its telecommunication networks enable market activity to take place across a vast and decentralized geographic terrain. In this sense, the NASDAQ network is located in multiple locations, from individual screens, to stories-high display screens in Times Square, to the massive server farms that collect and disperse data. Chapter 2 turns to the screen as a fossil figure, to examine the electronic market interface and to track the processes of dematerialization and automation that characterize electronic exchanges.

Chapter 3 investigates the locations and processes of electronic disposal and focuses on plastics as a fossil form and critical material that facilitates disposability. Electronics primarily consist of a complex composite of plastics, and plastics are the emblematic material of the "throwaway society." In this sense, plastics are both disposable and mobile, because once they are discarded, they also inevitably circulate through extended geographies. In the end, transportable electronic waste follows the path of the most undesirable forms of trash—from economically privileged country to poorer one. The primary exporter of electronic waste is the United States, a country that does not consider the export of waste to be illegal. But electronic wastes from the United Kingdom to Singapore turn up in places as distant as the rural districts and urban slums of China, India, and Nigeria. Recycling methods in these regions are typically toxic for both workers and the environment.[63] Chapter 3 trawls through these circuits in order to examine the material exchanges and geographies of disposal.

Chapter 4 considers electronic archives and memory as a site and fossil in which the accelerated temporalities of electronics become evident in sedimented form. The electronic archive operates as a kind of extended memory for the select electronic devices that are relegated not to the bin but, rather, to the archive and the museum. For every ton of electronic material cast out, a select portion ends up preserved in the halls of history. Much of the technology in the museum or archive of electronic history is inaccessible, however: ancient computers do not function, software manuals are unreadable to all but a few, spools of punch tape separate from decoding devices, keyboards and printers and peripherals have no point of attachment, and training films cannot be viewed. Artifacts meant to connect to systems now exist as hollow forms covered with dust. In this sense, the electronic archive can be seen as a "museum of failure."[64] It is a record of failed and outdated technologies. If it collects anything, it collects a record of obsolescence. The idleness of these electronic artifacts ultimately raises questions about how technology demarcates duration. How does one preserve media that have a built-in tendency toward their own termination?

Most electronics do not advance to preservation, however. Instead, idle machines, at end of life and end of utility, stack up in landfills, are burned, or are buried. More formally known in the Western world as the "sanitary landfill," the dump is the terminal site of decay, where electronics of all shapes and sizes commingle with banana peels and phone books. Plastic, lead, mercury, and cadmium break down and begin their terrestrial migrations. Electronics—media in the dump—require geological time spans to decompose. Chapter 5 begins and ends in the dump. Extending the discussions made in previous chapters, chapter 5 draws on the disposal practices developed in chapter 3 and the notions of time and preservation discussed in chapter 4. It dwells on the masses of scrap and peripherals, as fossil forms that are stripped, salvaged, burned, and finally dumped, often far from the sites of their initial consumption.

Digital Rubbish Theory

The dump is a site where objects typically absent of utility or value collect. Except through the work of invisible salvagers, from mice to treasure seekers, the material here is unrecoverable. Yet garbologist William Rathje suggests that the best way to investigate contemporary material culture is through this apparently useless garbage.[65] Much as archaeologists study the relics of the distant past, Rathje unearths the refuse of the recent past to measure human consumption. Garbology examines cultural phenomena by linking discarded artifacts with consumption patterns. Garbage Project crew members set out to landfills to draw core samples, tabulate and catalog discrete waste objects, and thereby chart significant patterns of consumption. In this sense, a dump is not just about waste, it is also about understanding our cultural and material metabolism. A dump registers the speed and voracity of consumption, the transience of objects and our relation with them, and the enduring materiality of those objects.

Electronics linger in the dump, where they stack up as a concrete register of consumption. The garbology of electronic waste may have an obvious reference point in landfills, but from Silicon Valley Superfund sites to recycling villages in China, there emerges an even more expansive array of waste sites where electronic debris expands, sifts, and settles. Electronics, media, landscapes, and waste are all linked and in constant transformation. From the virtual to the chemical and from the ephemeral to the disposable, the accumulation of these electronic wastes creates new residual ecologies and requires expanded practices of garbology. With electronic waste, it is possible to expand the thin surface of digital interfaces to encompass those material processes that work to support the appearance of immateriality. In the dump, our digital media and technologies turn out to be deeply material.

As the Garbage Project demonstrates, sorting trash into categories can become a habitual and absorbing project. A liminal zone, the in-between, the fringe, the outside of the inside, a site of expenditure and revitalization—the demarcations for waste are potentially endless. The ambiguity of determining when waste definitively becomes waste points to its role as a dynamic category. Waste oscillates in relation to ordering systems and structures of value. It is a variable within what Michael Thompson calls an "economy of values." As Thompson states in his authoritative *Rubbish Theory,* rubbish is a way of understanding the relative position of value relations.[66] Waste is, in this sense, what cultural theorist Walter Moser calls a "category of transition, a limit category."[67] Waste reveals the economies of value within digital technology that render valueless, for instance, a computer that is more than three years old. This collapse in value demonstrates assumptions within electronics—based on duration, novelty, and consistent consumption—that might otherwise go unnoticed, if it were not for the now-looming rubbish pile.

The interdisciplinary method of natural history developed in this book not only draws on studies of media, materiality, and technology, as already discussed, but also works through rubbish theories and waste studies, which critically inform this examination of the decay of electronics. The processes of materialization discussed here focus on "what was wasted"[68] in the manufacture, imagining, consumption, and disposal of electronics. The natural history method that emerges in this study is informed by these transformations and migrations to waste. Benjamin's method was, in fact, an early form of rubbish theory, where ruins, transience, fragments, and fossils served as key figures for thinking through exactly what is wasted in processes of materialization. The digital rubbish theory developed here weaves together these theories of waste and materiality in order to examine the material cultures and geographies of electronics through their dissolution.

Michel Serres asks, "Where do we put the dirt?"[69] Dirt, he suggests, may present another way for considering systems and relations through perceived imperfections. Where is the dirt of electronics? How does dirt inform the making of electronic materials and spaces? Electronic waste presents a crucial case study of dirt, of both how it is generated and where it is distributed. The nature of electronic waste suggests that it may be necessary to sort through the trash at an even finer scale to understand the implications of electronic modes of waste. Electronic waste, moreover, presents a critical subject for reevaluating our relationship with "new media." Digital technologies are disposable, and data is transient. Yet the rapidity of technological progress leads to enduring and toxic electronic materials. Electronic waste gives rise to a distinctly electronic version of garbology, a digital rubbish theory. Organized into chapters that focus on the previously described fossils and sites, the research that follows considers how remainders—and dirt—may be the most compelling devices for registering the transience of electronic technologies.

Silicon Valley Superfund site, hazardous waste log, 2005. (Photograph by author.)

Fry's Electronics Superstore, Silicon Valley, 2005. (Photograph by author.)

ONE

Silicon Elephants

THE TRANSFORMATIVE MATERIALITY OF MICROCHIPS

Out of the chip you can in fact untangle the entire planet, on which the subjects and objects are sedimented.

—DONNA HARAWAY, "Cyborgs, Coyotes, and Dogs"

Untangling the Chip

In Palo Alto, California, one can tune the TV set not just to the nightly news and game shows but also to local programming designed to instruct viewers on the finer points of computer systems. A computer system, one such program notes, is comprised of two elements: hardware and software. But here in Silicon Valley, it becomes apparent that the "system" of computing extends across a far wider horizon. In this sprawling landscape of sun and speed, one can detect other formations left over from the advancement of electronic technologies. Yet these formations inevitably fall outside the crisp diagrams that instruct on digital functions. Electronic technologies signal toward a future without residue, but in Silicon Valley, the epicenter of all things digital, one also finds the highest number of Superfund sites within the United States. Many of these sites, now in remediation, are saturated with chemical pollution not from heavy industry but, rather, from the manufacture of electronic components, primarily microchips. At one time, this part of California was founded on gold and the processing of gold ore. Now, however, this region is founded on another element and technology: silicon and microchips.[1] This is a region that grew out of silicon, that mineral bedrock of the digital age. Yet Silicon Valley is located here not necessarily for its wealth of raw materials (as silicon is one of the most abundant elements

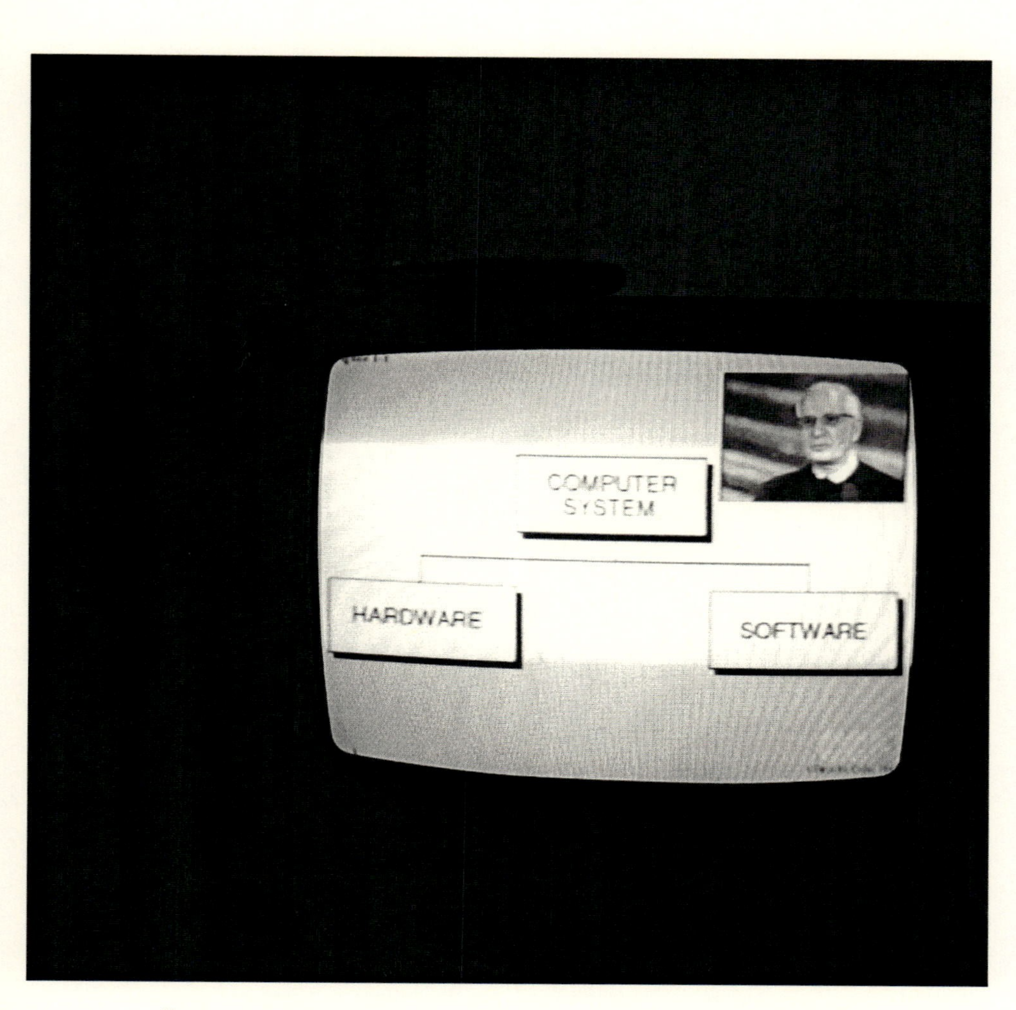

Computing systems instructional television program, Silicon Valley, 2005. (Photograph by author.)

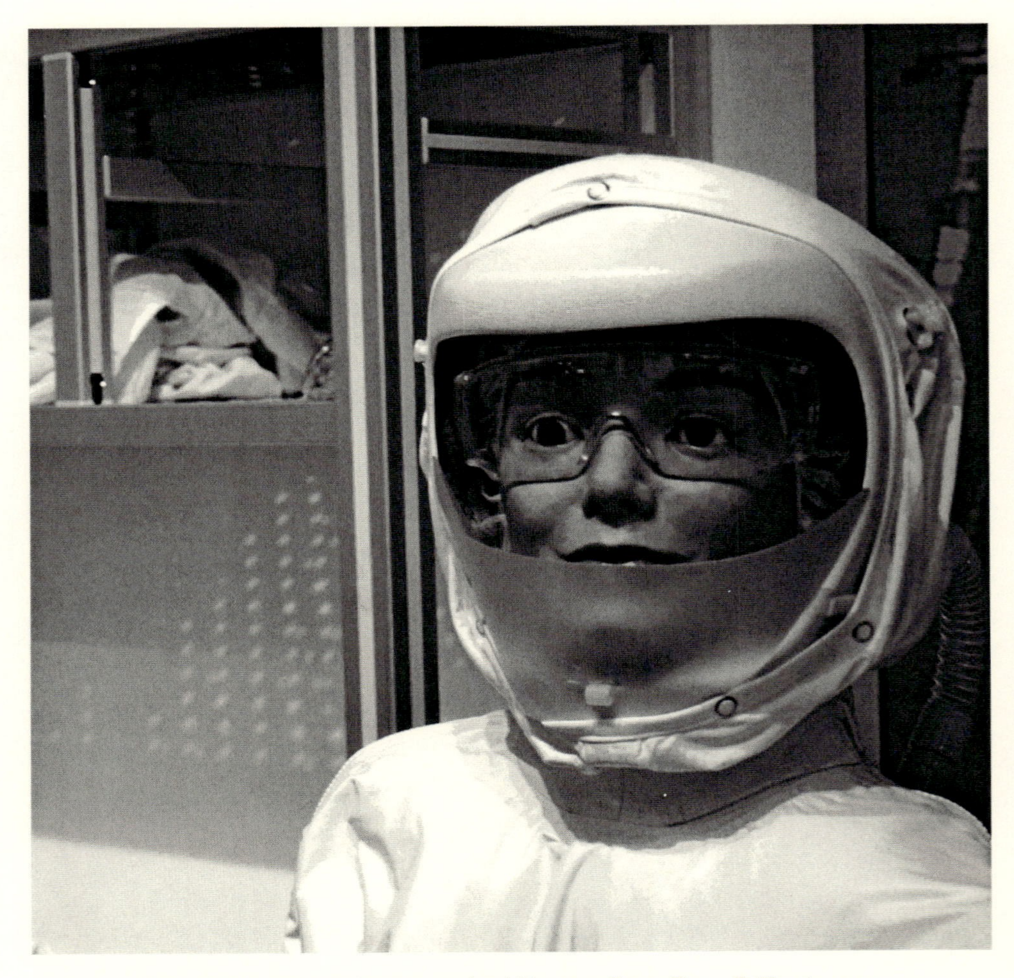

Model “fab” worker in bunny suit, Intel Museum, Santa Clara, California, 2005. (Photograph by author.)

Silicon ingot at Intel Museum, Santa Clara, California, 2005. (Photograph by author.)

anywhere in the earth's crust) but for its ability to transform silicon into microchips.

From silicon to microchip and from microchip to underground contamination, a complex set of mutations occurs to enable the development of electronic technologies. In the process of microchip manufacture, silicon does not long remain in its raw state but is transformed from ingots of silicon into thin wafers and finally into minute electrical assemblages. These assemblages, microchips, are the hardware that facilitates the transfer of information in the form of electrical signals, or on-off signals.[2] The transmission of information into bits, or binary units that correspond to electrical pulses, requires this composite of silicon, chemicals, metals, plastics, and energy.[3] It would be impossible to separate the zeros and ones of information from the firing of these electrical pulses and the processed silicon through which they course. A miniature device that performs seemingly immaterial operations, the chip, in fact, requires a wealth of material inputs.

This chapter "untangles" the chip by mapping the sites of its multiple transformations and by examining the residue that accumulates from these transformations. But microchips—and, by extension, information—have more than just an intricate material substrate in electricity and chemicals, and the scope of transformation from silicon to microchip is not limited just to the transfer of "raw" materials into pervasive electronics. Instead, silicon transforms from integrated circuits into electronic devices, chemical pollution and information overload, technological districts and architectural relics. The chip, as unearthed from manufacturing residues and dredged up in discarded devices, is embedded in complex material and cultural arrangements. By untangling this fossil, I do not arrive at a more discrete description of this technology but, rather, scratch the surface of a device that—despite its apparent simplicity and ubiquity—is exceptionally dense and entangled. I trace the extended contours of the chip in this way in order to begin to describe (though not quite) the "entire planet" in the enfolded layers of silicon and electrons, labor and new economies, contaminated bodies and environments, information and calculation, sprawling architectures and technological imaginaries. The material relations that can be traced through these contours—etched in the charged pathways and buried leftovers of electronics—sediment into this natural history of electronics.

To write this narrative, I do not comprehensively follow all the resource inputs and effects that are added and discarded in the process of chip manufacture;[4] instead, I select moments in the scattered relations of chip manufacture and information processing that resonate as key processes of materialization. This chapter traces the fossilized remains of the chip from manufacture and chemical inputs, through pervasive electronics and information overload, to end in the spatial arrangements and enduring material residues of Silicon Valley. Drawing on historical and contemporary reports of electronics and information technology, as well as descriptions of Superfund sites and the microchip production process, this chapter synthesizes the material and discursive aspects of microchips in order to describe a natural history of electronics that encompasses the abundant and the immaterial, the miniature and the toxic, the futuristic and the fossilized. These electronic proliferations fall outside the usual delineation of computing systems, but they are no less integral to how these technologies perform, materialize, and stack up as irrecoverable remainder.

The Chemistry of Speed

During the 1960s, Silicon Valley was home to a number of newly established technology firms that manufactured microchips, printed circuit boards, and developed related technologies that would transform both computers and electronics.[5] The same technology companies that were instrumental to the rise of electronics—from Fairchild Semiconductor to Intel, Raytheon, IBM, and Siemens—contributed over time to the formation of invisible Superfund sites through their widespread use of chemical compounds in the electronics manufacturing process. During and after production, many of these chemical compounds were stored in underground tanks made of metal and fiberglass. These tanks eventually leaked into the surrounding soil and groundwater. When the contamination was detected in the 1980s, it was revealed that tens of thousands of gallons of solvents had been leaking over a span of 10 to 20 years. Beneath the prosperous surface of Silicon Valley were plumes of poisoned groundwater that stretched over three miles long and 180 feet deep.[6] The removal of these underground contaminants continues to this day and may require several more decades of processing in order to reach acceptable levels of decontamination.[7] Chemicals that enabled the abundant manufacture and optimal functioning of microchips had contributed to intensive, long-term pollution.

The same basic process of microchip production prevalent at the start of the electronics industry is still in use today, and while the conditions for chemical transfer and storage have become less precarious, the manufacture of microchips still depends on a vast number of chemical compounds in order to assemble electronically charged devices. Microchip production may begin with the relatively benign and abundant material of silicon, but for silicon to be transformed into a conducting or insulating medium, it must first be chemically purified. This processed silicon is then melted and transformed into a silicon ingot, or rod-shaped piece of silicon, and sliced into thin wafers, the surface of which will be further altered through a chemical and material procedure of insulating and coating, masking, etching, adding layers, doping, creating contacts, adding metal, and completing the wafer. This elaborate and resource-intensive process transforms the conductivity of silicon and creates a grooved template. These charged pathways are the channels for the on-off electrical signals that will fire across and through the assemblage of copper transistors and chemically altered silicon. From this template, the wafers are cut into individual dies and packaged according to their use, from placement on circuit boards to insertion in other electronics, from mobile phones to calculators.[8]

From design to manufacture, the typical microchip (as produced at Intel) requires more than 200 workers, two years, and considerable material and chemical inputs to reach completion. The exact chip "recipe," as Intel terms it, depends on the particular use for the chip, but generally speaking, the input of chemicals, gas, light, and other materials can require up to 300 phases to reach a complete chip.[9] At each stage of the transformation of silicon on its way to microchips, a complex set of chemical and material inputs, together with considerable labor, contributes to the final chip. Many of these material inputs are not reflected in the end electronic product but are instead discarded as part of the hidden resource flows that contribute to electronics. In fact, microchips require far more resources than these miniature devices imply. To produce a two-gram memory microchip, 1.3 kilograms of fossil fuels and materials are required.[10] In this process, just a fraction of the material used to manufacture microchips is actually contained in the final product, with as much as 99 percent of materials used discarded during the production process.[11] Many of these discarded materials are chemicals—contaminating, inert, or even of unidentified levels of toxicity.

Chemicals are primarily used not just to adjust the electrical conductivity of the silicon wafers and to print or etch patterns onto the wafers where electrical circuitry will be placed but also to wash away any impurities or dust that may interfere with the functioning of the circuit. Dust can damage chips irreparably, wedging like boulders into the narrow pathways of transistors, gouging the thin architecture of chips and impeding the flow of electricity. The clean rooms within fabrication facilities (or "fabs") where microchips are assembled are zones specifically designed to be free of dust, as even the smallest impurity may ruin the minute transistors.[12] Ventilation systems, additional chemicals, ultraviolet light, and metal dust-free surfaces are required in order to achieve these contamination-free clean rooms.[13] Workers, moreover, don uniforms otherwise known as "bunny suits"—not so much to protect themselves from the chemicals but to protect the microchips from the dirt and debris that workers bring into the clean rooms. An uncanny inversion of waste occurs with microchip production, where clean rooms ensure the purity of electronics while simultaneously contributing to the contamination of workers' bodies, many of whom are low-paid immigrants and women of color. Indeed, it is increasingly suspected that the chemicals used in microchip manufacture cause everything from cancer to birth defects.[14]

The transformation of silicon into an essential material of the information revolution was in part enabled, as evidenced by the Silicon Valley Superfund sites, by an equally momentous revolution in chemicals.[15] Parallel to the electronics industry, there emerged multiple infrastructures and industries developed to supply, process, and dispose of the chemicals used in the manufacture of electronics.[16] In the space of 35 years during and following World War II, the development and manufacture of chemicals in the United States increased from nearly 1 billion pounds in 1940 to 300 billion pounds in 1976.[17] The postwar development of the chemical industry enabled rapid advancements in electronics. The increasing output of chemicals is closely paired with the development of electronics, and the introduction of new chemicals can even enable the basis for new electronic innovations. In this sense, the development of microchips not only depends on chemical compounds to ensure the accurate conductivity of silicon; it further depends on chemical compounds in order to increase conductivity. The terms of constant innovation and doubling of circuit capacity, which are captured by Moore's Law, have a chemical foundation.

In this unfolding material alchemy, it becomes apparent that a chemical revolution not only enables the information revolution by facilitating the transformation of silicon into charged integrated circuits; it further facilitates the abundance and speed of these technologies. The quicker the transmission required, the more highly processed the silicon must be.

Chemical transformations involve more than making apparently "raw" materials usable and efficient within processes of manufacture, however. These transformations are bound up with technological trajectories and imaginings, as well as arrangements of labor, economy, and resources, which together enable the profitable making of unprocessed silicon. Silicon becomes indispensable to Silicon Valley—and the information revolution—through these arrangements. While reference is often made to the quickening of information through digital technologies, it is evident that speed has a necessarily material, cultural, and chemical composition.

This natural history begins by tracing the remains from the seemingly most basic element of electronics in the form of silicon, but even this initial account makes perceptible how the transformation of silicon into chips enlists an entire inventory of material, environmental, bodily, technological, economic, and political effects. One material, one technology, untangles into entire fields of complexity. The density of a particular material, the histories and spaces within which it assembles, inevitably reveals interconnected narratives.[18] The recent history of electronics can also be read not just through silicon but also through plastics, metals, and any number of chemical compounds. Even with these compound material histories, we would further have to account for the fact that materials are, as cultural theorist Esther Leslie suggests in the context of coal and the Industrial Revolution, "transformative, transitory, non-eternal, productive."[19] To be productive, materials inevitably enter into processes of alteration, consumption, deformation, and decay. As materials are already tipping toward yet another process of transformation and exist only briefly in a seemingly absolute state, such as coal or silicon, this study of material histories raises the further question of why we should not tell such a history in reverse, by focusing on all that is wasted in the process of these materials coming into and lapsing out of finished and productive states. The material history of silicon and the microchip, that basic electronic component, exists not in an ideal or stable state but through multiple, migratory and transformative materializations. In the alchemy of electronics, silicon is transformed from a relatively common substance into a microchip and from a miniature electronic unit into a massive accumulation of waste.

Economies of Abundance

While the substances used in the manufacture of microchips contribute to pollution at both bodily and environmental levels, this condition intensifies with the sheer quantity of microchips manufactured. By the latest estimates, nearly 1 billion transistors "for every man, woman, and child on Earth" were set to be manufactured by the semiconductor industry; other estimates suggest that over 400 billion semiconductors have been manufactured worldwide to date.[20] Electronics manufacturing today is a leading market sector, a fact that is ultimately driven by the ongoing and global expansion of microchip applications. Even with reports of a transition to postindustrial or service economies having taken place in developed countries,[21] manufacturing is still vital to these economies, and this manufacturing is led by the electronics sector, which is one of the largest manufacturing industries—if not the largest—in the world.[22] The information economy is also a manufacturing economy. The appearance of light and resource-free information is, in fact, underpinned by the physical infrastructures of manufacturing. Indeed, one of the primary reasons that manufacturing may be relatively invisible has to do not with the elimination of this sector but, instead, with the offshoring of electronic production facilities. Silicon Valley may even enjoy the status of undergoing "cleanup" because so many of the potentially toxic production activities associated with electronics manufacture now take place in locations as far removed as Taiwan and Malaysia.[23]

As electronics become even more pervasive, the dilemma of how to contend with the chemicals and wasted materials that enable their production becomes even more pressing. It is hard to imagine how these miniature technologies can have such an accumulative and hazardous impact. But how did the chip become so pervasive? The microchip, or integrated circuit, developed in 1958, brought together previous advancements in transistors that would revolutionize the electronics industry.[24] Devices that once depended on bulky vacuum tubes to control the firing of electrical pulses could now run on a relatively minute and powerful assemblage of silicon, electrons, and bits. In the early 1970s, as the integrated circuit became even more sophisticated, it developed into the "computer on a chip," or microprocessor, which allowed for electrical control within a vast array of devices, from pocket calculators to microwave ovens to toys and automobiles. Microprocessors, or microchips, are now in such a wide range of products that personal computers count for only a fraction of all the microchips sold.[25] Indeed, when one enters the temple to electronics in Silicon Valley, Fry's Electronics in Sunnyvale, one sees just how many products depend on microchips for their functioning. Here are can openers and answering machines, irons and stereos, Web cams and toaster ovens, shavers and shredders. Serenading shoppers in this electronics superstore is even an automated player piano, which hammers out its anthems to the microchip.

The microchip may have depended on the abundance of silicon and chemicals for its manufacture, but an equally significant "invention" enabled the distribution and proliferation of the microchip. In the 1960s, one of the primary manufacturers of microchips, Fairchild Semiconductor, arrived at a basic strategy that would make microchips available for mass distribution. After significant cuts in military spending on electronics research, Fairchild sought to distribute its integrated circuits within the commercial market. Robert Noyce, then manager at Fairchild (and later cofounder of Intel), made the decision to sell its integrated circuits for less money than the devices actually cost.[26] Noyce calculated that by making integrated circuits pervasive and readily available, electronic products would eventually be redesigned to incorporate these superior and cheaper devices. By producing more integrated circuits, Noyce conjectured, the market would expand so that it would be possible to make a profit not necessarily through cost per unit but through volume and eventual necessity.[27]

Through volume, individual circuits would also increasingly cost less to manufacture and purchase. Instead of being a relatively expensive technology used primarily for military purposes, the integrated circuit became a technology available for mass application. Gordon Moore, Noyce's colleague at Fairchild, referred to this strategy of abundance as an "invention" that "established a new technology for the semiconductor industry," a technology in the form of markets.[28] These technologies and economies of abundance have a direct correlation to Moore's Law, penned by Gordon Moore in a paper originally forecasting that the number of transistors on integrated circuits would double every 18 to 24 months and thereby effectively double processing speed. This law has become a nearly inviolable principle for the rate of electronics advancement.[29] Moore's Law constitutes a code and duration for continual increases in the speed of processing. It is the technological instantiation of ongoing and even exponential growth. Such growth inevitably has material and informational dimensions, as the doubling of capacity translates into more chemicals, more devices, and more information—and more waste.[30] In the Intel Museum in Santa Clara, California, a "Microprocessor Hall of Fame" records these steady advances in the form of historic, outdated chips that document the decreasing size and increasing capabilities of computing power. Spanning from the 4004,

Intel's first microprocessor, developed in 1971 with only 2,300 transistors, to the Intel Itanium processor, developed in 2005 with over 1 billion transistors, these chips gleam as fossilized remains, bearing the inscriptions of technological advance.

This shift toward volume as well as steady advances in processing speed, the "invention" of a technology and economy of abundance, helped to make Fairchild (and subsequently Intel) a primary producer of microchips.[31] Both Noyce's strategy to saturate markets with microchips in order to allow an emerging technology to take hold and Moore's reference to this strategy as a technology or invention in its own right suggest that new economies emerge concomitant with new technologies.[32] New economies, together with new arrangements of labor, altered material and chemical inputs, and spatial distributions, help to create the very conditions through which a technology can take hold, persist, and even become seemingly natural. The emergence of these new economies and related infrastructures requires more than the deliberate intentions of actors or inventors, however, and as I suggest in this study, these material sedimentations can perform in unexpected ways, particularly as they accumulate toward conditions of waste and overload. Furthermore, the imaginings of and strategic discourses describing pervasiveness are as crucial to understanding the processes of microchip development and materialization as is the emergence of economies that enable such pervasiveness.

Pervasiveness of microchips—and, by extension, pervasiveness of electronic devices—was then part of the design and imagining of chips, and this was not just so at Fairchild and Intel. In 1964, Patrick Haggerty, head of Texas Instruments at the time, forecast that electronics would become completely common "if the vexing technical problems related to reliability and containment of fabrication costs are overcome."[33] By successfully overcoming technical constraints, it would be possible to achieve a positive feedback loop where electronics contributed to their own proliferation. This is what Haggerty referred to as "the ultimate pervasive character of electronics," where electronics would become so ordinary that there would be no aspect of society that was not in some way informed by electronic processing.[34] Indeed, as this study notes, microchips of all types are now embedded in everything from computers to consumer electronics to control systems.

The miniature microchip developed in a brief period in the 1960s to the early 1970s, to emerge in the mass quantities and pervasive uses common today. But the pervasiveness of electronics occurs not just through material resources, chemical revolutions, cheap labor, the mass distribution of microchips, or even the lack of technical constraints; it also takes hold through the conversion of nearly everything, from media to human memory, into information. Electronic information technologies facilitate the digitalization of a vast array of media and data, such that almost anything can now be accounted for and transmitted in terms of informational bits, or zeros and ones. The proliferation of microchips, in other words, correlates with the proliferation of information. Yet both forms of proliferation have corresponding forms of waste, from chemical contamination to information overload. Strangely enough, because the waste generated from microchips is so often invisible, it is perhaps through information overload—a seemingly more immaterial condition—that we can begin to gauge the complex transformations that accompany digitalization.

Digitalization: The Midas Touch

When engineers and mathematicians Claude Shannon and Warren Weaver wrote their classic text, *The Mathematical Theory of Communication,* on information theory in 1949, they were concerned with defining a measure of information—the binary digit, or bit, that could be readily used within electrical devices.[35] Cautioning that their definition—which encoded information through the on-off pulsing of electrical signals—should only apply to limited technical situations, their model instead came to serve as the predominant interpretation of information as a unit free from meaning and context.[36] From this model of efficient and all-encompassing information, nearly everything came to be rendered in terms of information, from organism to economy. As though under the spell of Midas, who had the mythic ability to turn anything he touched into gold,[37] digital devices have the ability to transform anything encountered into some register of information. The management of information through digitalization establishes a standard medium and mode of measure, with an extensive capacity of assimilation. From sensation to speech, information, as a universal standard, in some ways even constitutes a new currency.[38] With nearly everything now rendered in terms of information, the question is whether anything actually falls outside of information, or is undigitizable. What are the limits to digital absorption, and what is the fallout from such complete assimilation?

The bit, as defined by Shannon and Weaver, is an ideal communication device and strategy of control and efficiency. It maximizes channel capacity and the speed of communication. But this use of information has a longer history, where information has repeatedly been used as a device to control conditions of overload. Indeed, the "information society" emerged, as communications scholar James Beniger argues, "in response to the nineteenth-century crisis of control."[39] The accelerated rates of production that arose with mechanized industry brought about a rising need to manage production, monitor supply and consumption, and coordinate distribution. Information and communication were central to establishing control over increased production and became strategies for coordinating and distributing goods and monitoring labor.[40]

The management of information involves the application of technologies that control yet contribute to the problem of proliferation. The threat of overload can give rise to adaptation and innovation, where new technologies are required to trawl through all the new data. In the loop from crisis of proliferation to crisis of control, excess data gives way to technologies for managing that excess.[41] Electronics could be located in this loop, as technologies that, on one level, improved the efficiency of calculation and communication. Yet these technologies also operate as technologies of excess; they are the very devices through which we can trace emerging forms of proliferation. This is the dilemma of information, where the line between information and entropy is a thin one.[42] Information technologies contribute to the very proliferation they attempt to manage.[43]

The flood of information is both a consequence of and contributor to the pervasiveness of electronics. From speculation about what could be achieved through widespread use of electronics in the 1960s, to the introduction of the first integrated circuit used in the pocket calculator in 1971, to the development and increasing use of home and office desktop computers in the 1980s, the proliferation of these devices arose parallel with new languages—and even philosophies—for analyzing electronic information. The pervasiveness of electronic technologies may even contribute, through sheer quantity, to changes in the movement and definition of knowledge. In language that could be read as symptomatic of its subject, Jean François Lyotard suggested in his government-commissioned study on information, *The Postmodern Condition: A Report on Knowledge,* that "the proliferation of information-processing machines" would contribute to changes in the "circulation of learning" and in what counts as "knowledge statements."[44] In Lyotard's assessment, the amount of information and its devices in circulation could transform cultures of knowledge. More information may require more technologies of storage and processing. Proliferation becomes a structuring and dynamic feature of information and electronics, and similarly informs the language and theories used to assess the effects of these new technologies. Such a transformation does not just occur at the level of structural definitions of information and knowledge but also involves new materialities and technologies for processing large quantities of information in order to generate knowledge.

The terms *overload* and *explosion* often emerge in attempts to capture the increase in information that has accompanied the burgeoning industry in microchips and the increasingly pervasive presence of electronics. Any number of studies refers to "data smog" or "communications glut" to describe the increase of digital devices and media.[45] The material and discursive features of electronics are intertwined, so that technologies of volume become inseparable from the language of volume. Rhetoric—as much as hardware—becomes a critical type of fossil to collect and study in this natural history. Recent reports on the status of information reveal the extent of this discourse, where attempts to calculate the growth in digital technologies and communications media arrive at estimates that are nothing less than exponential.

In this light, a study titled "How Much Information"—initially conducted during 2000–2003 at the University of California, Berkeley, then later based, for updating and revision, at the University of California, San Diego—intended to calculate and measure the breadth and depth of the digitally induced explosion by documenting increases in media and information.[46] As its title suggests, the report strove to measure not just information but also its apparent boundlessness. In attempting to assess the scope of information-based growth, the authors of this study arrived at a methodology that intended to "measure only the volume of information, not the quality of information in a given format or its utility for different purposes."[47] Because all media contain some aspect of "information," a common standard of measurement is necessary in order to tabulate all extant information. Since most new information is created, transmitted, and stored in digital format—what the study's authors refer to as the "dominance of the digital"—the authors decided that digital measurements would be the best gauge of this explosion.

Digital measurement appears to be the best means to capture the information explosion; yet it is possible to extend this measurement one step further. The authors of "How Much Information" decided that terabytes would be the ideal unit for measuring information. Terabytes are useful not just because they are a digital mode of measure but also because they are an abundant degree of measure—1,000,000,000,000 bytes, or a thousand times more bytes than a gigabyte. Yet all the terabytes of new information each year require yet another standard of composite measure, the exabyte—1,000,000,000,000,000,000 bytes. The scale of these digital measurements captures the incredible volume of information produced annually. Yet at the same time that terabytes and exabytes aid in recording the volume of information, digital devices continue to enable even new levels of voluminous production, storage, and transmission. To take just one example, "How Much Information" finds that the number of photographs taken in any given year is estimated to be more than 80 billion. With the aid of digital cameras, image phones, and various instruments of duplication, the capture and transfer into digital format is instantaneous. Working within these digital measures, the authors estimate that all "information stored on paper, film, optical, and magnetic media totals about 5 exabytes of new information each year."[48] Yet stored information is only a fraction—one-third—of all information that is in circulation, whether in the form of telephone or Internet transmissions, which the authors estimate totals 17.7 exabytes annually.[49]

Everything in the air, over the wires, stacked up in libraries, or collected on home digital cameras becomes a potential source of new information to be measured. "Brand-new" information also features as worthy of measure, but in this sense, the aforementioned study does not consider how much information may be redundant or duplicated. The preoccupation with measuring volumes of information flattens existing media and diversity of formats and makes information "new" simply through the acts of digital translation and measurement. The digital is central not just for its new production and ease of measurement but also for the conversions that it allows—namely, that everything can be captured within that universal machine, the digital format. To compute is to calculate. The pairing of digital technologies with enhanced powers of measurement and calculation is more than a technological advance; it informs the very operation of these technologies.[50] An attempt to control and manage a digital explosion through measurement, "How Much Information" contributes to that explosion through its inevitable bias toward calculation. In attempting to capture the volume of information growth, the study conveys the quality of quantity, where the self-reinforcing and accumulative tendencies of calculation contribute to altered organizational and material arrangements. With such strategies of measurement, excess calculation may further give way to new qualities and standards of measurement, where calculation enables ever-shifting, rematerializing practices.[51]

With the "dominance of the digital," there is a tendency not just to calculate to the point of excess but also to compress more media and material into compact digital formats. Compression does not just consist of minimizing file sizes and lowering resolution of already digital media; it also involves shearing off the unwieldy and bulky aspects of less-compact media, from volumes of books to reams of paper and reels of film. As the authors of "How Much Information" remark, the "common standard of comparison" used to assess media types also involves the problem of determining a comparable "level of compression" across media formats, where the resolution of a book might correspond equally to the resolution of a telephone conversation. Through the levels of compression used in their study, "a small novel" becomes equivalent to one megabyte of information, while "a pickup truck filled with books" compares to one gigabyte of information. At the top end of the spectrum, "all words ever spoken by human beings" equal five exabytes.[52] Pickup trucks, moving vans, and entire libraries condense into digital formats of storage and measurement; yet through compression, an enormous rise in information volumes occurs. As of 2003, the last year the "How Much Information" study was updated, estimates of total new information produced each year reached over 22 exabytes. The study set out to estimate the amount of information produced annually and found that figures in 1999 were obsolete by 2003. In this sense, the study occupies a transitory position, as its findings must be constantly updated in order to capture just how rapidly information is growing. What these constantly renewed numbers reveal is just how difficult it is to measure—to classify and stabilize—the information explosion. This is an explosion that we are compelled to measure and contain because digital devices seem an ideal technology for ordering media through calculation. Yet digital devices appear to contribute to the very explosion they measure.

The flights of numerical imagining that digital technologies enable have more than a recent history, however. In his classic 1945 text "As We May Think," computing pioneer Vannevar Bush discusses the possibilities for collapsing media, such as film and books, to a miniscule size with technologies of compression. Bush proposed a technology that took the form of what he called the "Memex," a technique for compressing and accessing vast stores of information. In the process of elaborating on the benefits of technologies of compression, he writes,

> The *Encyclopedia Britannica* could be reduced to the volume of a matchbox. A library of a million volumes could be compressed into one end of a desk. If the human race has produced since the invention of movable type a total record, in the form of magazines, newspapers, books, tracts, advertising blurbs, correspondence, having a volume corresponding to a billion books, the whole affair, assembled and compressed, could be lugged off in a moving van.[53]

Bush describes an economy of scale—compression—that moves parallel to an economy of abundance. Instead of minute technologies accumulating toward the saturation point, this is an inverse form of saturation that takes place through the compression of information to its most minute form. A cost-saving technique, a mode of measurement, and, more commonly today, a mode of preservation, compression also makes room for more information to be generated. As Bush notes in his essay, these technologies of compression allow the most information to be stored and transmitted efficiently, which enables increased production and distribution. The compression and storage of millions of bits of information ultimately allows for the production of billions.[54]

Compression establishes the scale of implosion, which differs from explosion in that it reorders the qualities of an already saturated medium or situation. Saturation, a rushing inward rather than just a dispersing outward to occupy distant terrain, aptly characterizes this era of electric intensity. The growth of media, the condition of overload, is as much a media implosion as a media explosion. Implosion is "compressional." It is involving, rather than enlarging or expansive.[55] Implosion is no longer a question of extending to the unknown edges; it is amplifying the intensity of the already mapped. With implosion, media and material are worked and reworked, concentrated and differentiated. Compressing all media into a standard but proliferating unit of information involves removing those material structures and spaces that may, in the end, have facilitated the very way in which we access that information.

Often, with intense quantities of information, seemingly more archaic material structures allow for the easiest access and transmission of vast stores of digital information. Because rates of digital transmission still lag well behind the quantity of information that can be stored and generated on digital devices, computer scientists are known to mail entire hard drives through the post, as it is a more efficient way to deliver terabytes of information.[56] The return to palpable and even predigital material structures to carry digital data is in many ways implosive. This is an informational world that has been entered into digital format; its expanses have been charted and captured. But the means of accessing and transmitting data occurs by reinserting that data into existing physical infrastructures. Calculation not only may be self-reinforcing—the quality of quantity—but may also create other material arrangements and relationships that emerge through, but also exceed, devices of measurement.[57]

While this discussion of information overload may seem remote from the unwieldy and extensive remainders of electronic waste in the form of abandoned computers and other discarded electronics, it is, in fact, an integral part of the processes of electronic materialization. The imagining of relatively malleable and even immaterial structures of information could, in many ways, be seen to enable proliferation and to set in play economies of abundance for which resources and labor appear to be of little consequence.[58] The proliferation of information informs material processes. Abundant information requires electronic devices and chemicals, information economies and landscapes. In the last section of this chapter, I return to the landscape of Silicon Valley to consider the extent of these material infrastructures that keep in motion so many imaginary moving vans of substance-free information.

Environmental Overload

Overload-informed material transformations span from the proliferation of microchips, to the apparent immateriality of excess information, to the spread of technological districts. The vestiges from silicon transformation are to be found not just in the form of bits but also at the scale of landscapes, such as Silicon Valley. Overload, moreover, is a condition that not only afflicts information but also is relevant to environments. As a concept, overload initially meant conditions of environmental excess.[59] Today, environmental overload might include not just excessive (urban) stimuli but also ecosystems at capacity, landscapes marked by saturated soil and groundwater, and sites of maximum economic development and accumulation. In the same way that informational overload is paired with continual strategies for contending with proliferation, environmental overload exists alongside accompanying strategies for dealing with saturation.

Silicon Valley is a landscape so contingent on digital technologies that it could almost appear to be a "virtual geography"[60]—an environment informed as much by the imagining of and through digital technologies as it is by their actual manufacture and development. But this relationship between the digital and the geographic again reveals not the elimination of spatial or material resources but, rather, the distinct material inscriptions and geographic arrangements that occur in landscapes oriented toward the development of electronic technology. In my mapping of the 29 Silicon Valley Superfund sites, those residual spaces from microchip production, I crossed the trail of interconnected microchip fabs and recyclers of chemical barrels and drums. EPA plans and sections detail the extent, both in time and space, of the chemical spread, across decades and into the aquifer. In among these sites, wasted relics of chemical barrels and electronic appliances shore up Silicon Valley parking lots. Netscape Headquarters is a model project in this collection of sites, the location of a successful remediation cleanup from previous pollution by Fairchild Semiconductor.[61] Scattered within this space and visible across the intricate freeway exchanges (six to eight lanes of dense traffic) are vast sprawling parking lots marked with corporate logos, in some cases tens of meters tall—Adobe, Intel, Yahoo! Aerial images of this landscape indicate landfills and salt evaporation ponds, a savanna-edge landscape that is characterized by patches of brown and irrigated green. At the street level, there are miles of spaceship-shaped office buildings, palm trees, turf grass, and asphalt, scattered together with mini-malls, fast-food restaurants, and chain hotels with virtual blue swimming pools. Bungalows house the working class and millionaires alike, albeit at radically different prices, depending on the location of the real estate.[62] Silicon Valley is an extensive, developed, and resource-intensive environment. Information, in all its fleeting immateriality, bears a direct relationship to this landscape.

In order for technological development and economic accumulation to take place, they must be located in and bear relationships to places.[63] Silicon Valley is such a landscape, a conglomeration of silicon wafer fabs and freeway circuits, research labs and chemical suppliers, its infrastructures built up for the purpose of accumulating the resources necessary for "digital dominance." These spatial infrastructures are not ancillary to the information revolution. They are, in fact, critical material resources and relationships within the dynamics of economic growth.[64] Silicon Valley, a landscape geared toward digital production, is built not just from virtual bits but also from sand and asphalt. Proliferation and material transience inform the qualities of information as well as landscapes. Silicon Valley houses freeway circuits and office complexes that spread across this region. Yet these same structures may be subject to removal or modification, whether through new economic development or the need for environmental cleanup.[65] Silicon Valley may have engineered its own geology, where the longer durations of environmental processes have quickened to a digital pace. Architecture, urban development, and transport emerge and subside on time scales that approximate a more electronic register of materiality. At the same time, we can imagine archaeologists trawling through this landscape hundreds of years from now, uncovering chemical barrels and electronic appliances, silted under asphalt and turf grass.

This tension between rapid development, apparently weightless technology, and the denser materiality of environments plays out across more landscapes than just Silicon Valley. Such speed and ephemerality of environments and technologies are relevant not just within the vague boundaries of Silicon Valley but also on a global scale. Silicon Valley is a model "postindustrial" landscape that multiplies, a space of technological purpose replicated from the Silicon Fen in the United Kingdom to the Silicon Mountain in Colorado, in addition to the digital cities and digital zones found everywhere from Seoul to Dubai.[66] The expansion of districts patterned after Silicon Valley demonstrates the global impact of this particular landscape. With the increasing tendency to outsource and offshore the manufacture of digital technologies, the relationship and impact of Silicon Valley magnifies from wafer fabs to technology parks. With Silicon Valley, a space of multiple remainders, waste does not just linger on the periphery but, rather, is integral to centers of production and to the dynamics of economic growth worldwide.

The system of hardware and software thought to contain digital systems breaks open once again to reveal intersections with other landscapes. Rattling around the edges of these apparently discrete systems are the residues from electronic and informational proliferation and from ongoing spatial, economic, and technological development. Beyond the making and the clearing, the proliferation and control, the boom and the bust, are remainders that suggest other narratives for describing the material and imaginary aspects of electronic technology. What this remainder reveals is that digital technologies do not oscillate exclusively between control and proliferation. The stack of material discards left over from the manufacture and decay of these technologies suggests at least this much. This surplus has an unacknowledged impact on these systems, an impact that cannot be completely encapsulated, because it is so unpredictable. Remainder is more than an opposing pole. It does not play the role of inversion. It is irreducible. Taking up this point, Jean Baudrillard—sounding more like a protoenvironmentalist than a postmodern philosopher—elaborates, "It is no longer a political economy of production that directs us, but an economic politics of reproduction, of recycling—ecology and pollution—a political economy of the remainder."[67] Remainder breaks with sustained cycles of production; it moves us past what might be seen as a Marxian concern with the way raw materials are mobilized for production. The practices and materialities of recycling and remainder cannot be fully reincorporated, and so, through their intractability, they give rise to changed ecologies and economies. Interfering with any notion of a simple feedback loop from production to consumption, remainder calls attention to the aftereffects and transformative material arrangements that emerge through the density of our technological and cultural practices.

Electronic remainders guide us toward a narration of technology that is oriented not necessarily toward production or control or toward progress and great inventors. Instead, they compel us to describe these technologies and their residues from the ground up, by describing their material traces and entanglements, or, in the case of Superfund sites in Silicon Valley, from the ground down, by digging into those deep sedimentary layers thick with the residues that accumulate into a natural history of electronics. The material in this chapter describes the ways in which the contours of the chip untangle into Superfund sites in Silicon Valley, bodily contamination, pervasive electronics, information overload, and environmental and architectural remainder. These conditions together begin to describe a natural history of electronics that at once captures the "naturalized" narratives of technological advance as well as the natural-cultural residues of technological production. As I suggested at the beginning of this chapter, the initial transformation of silicon that begins the process of microchip production does not long remain within an ideal and stable state. As silicon passes through a number of migratory states, it quickly unfolds into a mass of other materials, economies, and spaces required for its transformation and deformation. Rather than focus exclusively on the initial promise or assumed progress of digital technology, I focus on this technology's remainders, to better understand what other stories and orders of experience these unruly materialities and operations generate.

In this chapter, I have worked through the sedimented natural history of silicon from the perspective of residue and remainder, in order to untangle the (fossilized) chip and arrive at a more complex set of inputs and effects. As signaled in the introduction, the conditions that have emerged as effects of electronic materiality extend well beyond the *production* of digital technologies and control of information described in this chapter. In the next chapter, I focus on another related type of fossil—the electronic screen—and investigate the ways in which strategies of dematerialization and proliferation, as discussed here, are processes of materialization that enable and characterize electronic transactions. Although the electronic interface appears to be dematerialized and even weightless, it, too, is bound up with material transformations and remainder. These remainders, however, surface neither in spaces deep underground nor in the density of information but, rather, in the orders of electronic time and exchange that emerge at the interface of electronic markets.

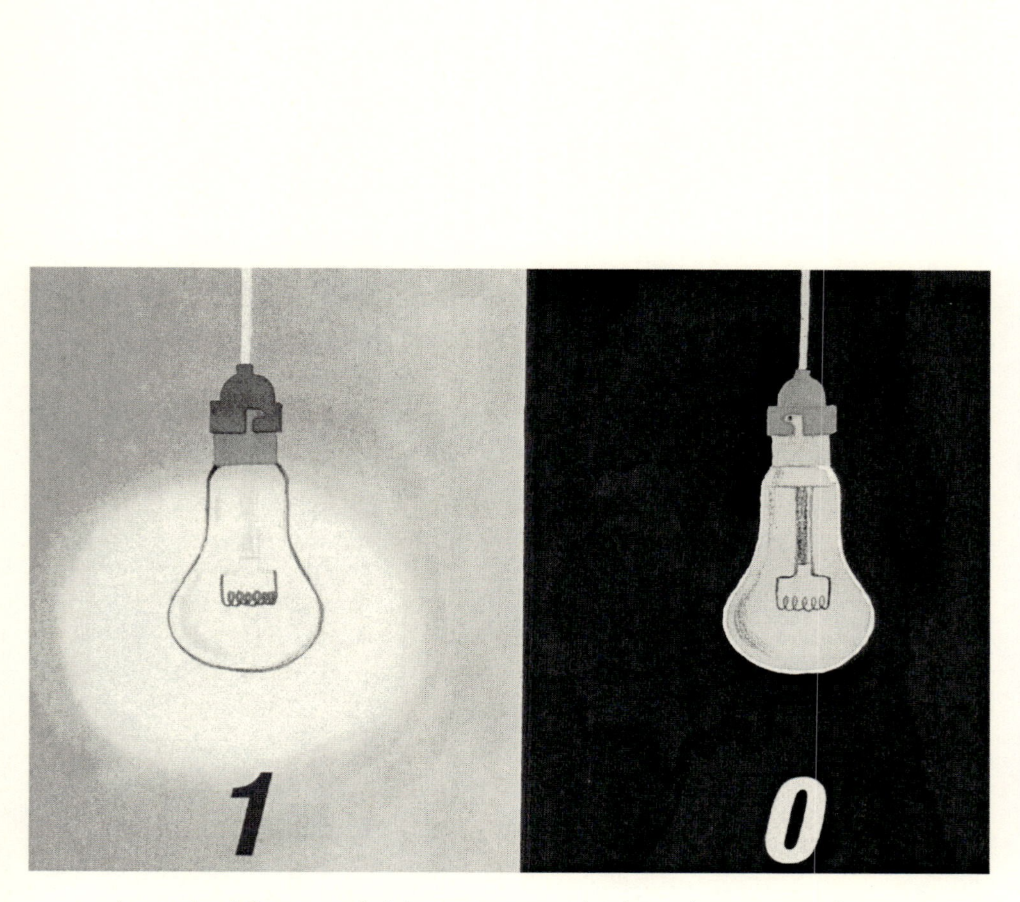

International Computers Ltd. instruction material on binary logic, ca. 1970, Science Museum of London. (Courtesy of Fujitsu.)

Silicon Valley, vacant buildings, 2005. (Photograph by author.)

TWO

Ephemeral Screens

EXCHANGE AT THE INTERFACE

Our best machines are made of sunshine; they are all light and clean because they are nothing but signals, electromagnetic waves, a section of a spectrum, and these machines are eminently portable, mobile—a matter of immense human pain in Detroit and Singapore. People are nowhere near so fluid, being both material and opaque. Cyborgs are ether, quintessence.

—DONNA HARAWAY, "A Cyborg Manifesto"

The signal and the thing are not as cut off from each other as they say.

—MICHEL SERRES, *The Parasite*

Stock Exchange: Removing and Transplanting

Throughout most of its history, the New York Stock Exchange (NYSE) conducted the majority of its transactions through the medium of paper. Ticker-tape remainders and other paper scrap that recorded the latest stock quotations circulated and accumulated in the flurry of trading. At the close of each day, the trading floor would be littered with these left-over papers that had been fleeting carriers in the circuit of exchange. In 1968, the artist Dennis Oppenheim collected some four tons of this "paper data" from the floor of the stock exchange and relocated it to a rooftop in New York City.[1] His project, *Removal Transplant—New York Stock Exchange,* transferred and revealed the ephemeral material that was expended in the process of market transactions. Transactions between distant places were recorded on the paper scraps, but the rapid pace of exchange had rendered the printed matter residual. By placing the litter against the Manhattan skyline, Oppenheim moved the overlooked

NASDAQ MarketSite, view from Times Square, ca. 2004. (Photograph courtesy of NASDAQ.)

debris of communications to the foreground, demonstrating that it, too, was an essential building block that cut through the core of the city.

This "removal" and "transplant" describes the movement and relative abstraction of material within markets—not only paper material, but also the material of commodities. Each scrap of paper and ticker tape tracks a record of the fluctuations of the market, as well as the values of the exchanges and commodities it represents. This is a core operation that stock exchanges perform in the distancing and removal from materials and sites of production. In addition to the process of removal, such movement enacts a displacement and transformation: through the process of exchange, objects materialize and dematerialize based on their value and the contexts in which they circulate. Exchange is a process of removing, transplanting, and transforming. In this way, another kind of removal has since occurred in the spaces of trading. With the arrival of electronic telecommunication networks and computing systems, the paper ticker tape has transformed into screen-based displays, and the exchange floor has migrated from a central physical location to a dispersed network of millions of terminals scattered across the globe. Now, markets have been removed and transplanted to electronic screens and networks. With this removal, even the stock exchange appears to have dematerialized.

This chapter focuses on the electronic screen as a space and device through which "the signal and the thing" (the subject of the quote from Michel Serres at the beginning of this chapter) seem to be disconnected and dematerialized. Looking specifically at the deployment of screens in and through the electronic network of NASDAQ, I consider how screens within this particular market are critical objects through which to examine processes of materialization and dematerialization, specifically through exchange and the rise and fall of value. By articulating this relationship between signal and thing, my point is not to draw a direct and unproblematic line between these but, instead, to discuss how the signal and the thing are bound into shared material processes. From screens to networks and from networks to software and automation, technologies and programs have emerged for distributing and mobilizing matter—the thing—toward signals and light. Through my analysis of NASDAQ and by considering the screens that form a considerable part of the traffic in electronic waste, I excavate the layers and processes of materiality that sediment through electronic exchange, screen imaginaries, and the performativity of networks and software. These processes suggest that the signal and thing are co-constitutive, that commodities, value, and matter emerge or dissipate not through the sheer inertness of things but through the processes that allow these things to cohere—even if momentarily. Similar to the chip, the screen of electronic markets is another key site and fossil from which to rematerialize electronics in order to assemble a dense material record—a natural history—of these devices.

Electronic Performativity

The introduction of electronic trading occurred primarily through the National Association of Securities Dealers Automated Quotations system, which began to electronically display stock quotes through an automated system in 1971. The display system could be accessed by traders throughout the United States, almost in real time. With this electronic network, NASDAQ dispensed with a central exchange floor and instead established a dispersed market that spanned the entire United States. In many respects, by decentralizing and distributing its trading across electronic networks, NASDAQ achieved greater coverage and, eventually, greater volume of exchanges. It previously played a secondary role to the authoritative NYSE, but through its electronic network, NASDAQ became recognized as the world's first electronic stock market. This network now makes real-time stock quotes available and allows for order execution from over two million terminals worldwide, while at the same time making delayed market information available on its Web site.

NASDAQ registers not just the electronicization of markets but also the rise of information and communication technology values within markets, as it has come to be known as an index representing a high proportion of technology companies.[2] Microsoft, Intel, and Google all feature on NASDAQ. The efficiency and speed with which trading may be conducted and the fact that NASDAQ is "the largest electronic screen-based equity securities market in the United States," listing over 3,250 companies, mean that it "trades more shares per day than any other U.S. market."[3] In its electronic transactions and as the emblematic index for the "new economy,"[4] NASDAQ has moved from a previously novel electronic display system to become a mode of exchange that influences dynamics of value and devaluation, as well as materialization and dematerialization.

March 10, 2000, is well known by now as the date when NASDAQ reached its peak but also experienced a sudden plummet in value. From

5,048 points, NASDAQ quickly fell 3,000 points. It had lost 60 percent of its value by March 2001 and continued to decline. While this crash seemed to portend the end of digital dominance, it has, in many ways, had a contrary effect. The value of NASDAQ now oscillates below its historic peak, yet its volume and speed of trading has continued to accelerate. With this performance, however, the "new economy" was more continuous with historic economic practices than the term suggested. Speculative bubbles, fueled by technology and the promises in new development that technology generates, are a long-standing dynamic within stock markets.[5] Financial crashes and the waste generated through these crashes, whether in the form of devalued stock or ruined companies, can be a crucial dynamic in the generation of value. This is because rubbish is a generative dynamic.[6] Waste—the possibility of devaluation—also enables the opportunity for revaluation. Value is unstable; things move through stages of value and may in all likelihood become waste. The generative dynamic of waste, then, describes a possible limit of value as much as a condition for potential recuperation of value.[7] Yet this same set of dynamics translates into discards, from obsolete devices to devalued shares to bankrupt companies. Indeed, overvalued Internet companies were not the only casualties of the dot-com crash. As the previous chapter notes, the material remainders from this rapid devaluation can be discovered in the vacant buildings and empty parking lots that periodically litter the landscape of Silicon Valley. Electronic commerce has more than a passing connection to electronic waste. As the present chapter suggests, these cycles of value are not without remainder.

In addition to these wavering cycles of valuation and devaluation, NASDAQ has achieved a more thoroughgoing effect on markets through its electronic network and through the speed, volume, numerical precision, and automation that underpin this network. Setting the pace, as it does, NASDAQ transactions are informed by the temporality of digital technology. This is a timing that is bound up with the instantaneity and mutability of turning profits on the tick of the virtual ticker tape.[8] Yet the electronic market sets the pace in more ways than one, for the speed of trading has as much to do with the rise and fall in value as it does with the accelerated movement and programming of exchanges. These electronic markets are bound up with performative registers—material, temporal, and rhetorical deployments that involve affective as much as calculative maneuvers. In fact, the calculative becomes inseparable from the performative.[9] But such performativity is, as suggested throughout this study, often unruly. In a volatile market, the inevitable devaluation and destabilization of commodities, share prices, and futures can potentially move at an even faster pace through the enhanced calculability afforded by electronic exchanges. As was revealed by the losses from the crises involving subprime mortgages and credit in 2007, such calculations can contribute to even more complex entanglements and oscillations of value.

Electronic markets emerge through the material and performative qualities of digital technologies at the interface and through the extended effects of these machines that, as wryly suggested by Haraway in her quote at the beginning of this chapter, are seemingly comprised only of sunshine and signals. NASDAQ is more than a financial instrument. It sets into play a performative and material economy, which has political, cultural, and environmental effects. From the speed and volume of exchanges, to the volatility of values, to the apparent "removal" of material structures, this electronic market contributes to the circulation, dematerialization, and devaluation of electronic technologies. The performativity of NASDAQ can even create conditions of "counterperformativity," where the failure of market devices can interrupt their performance.[10] Electronic markets perform in ways that exceed expectation: they reach saturation, collapse, generate waste, and recuperate, sometimes almost instantaneously. The performative or "expressive" failure of markets suggests that these processes of valuation and materialization involve something more complex than rational, calculative intention.[11]

NASDAQ does not wholly encapsulate the extent or force of the new economy. In fact, electronic market structures and digital technologies are more pervasive, complex, and unpredictable than a single index can measure.[12] The electronic, captured in the eponymous prefix *e-*, includes, as media theorist Rita Raley writes, "communicative networks, electronic commerce, modes of production, and global financial markets."[13] The whole of market activity cannot be explained through a discussion of the materiality of electronic exchange. Yet in many ways, electronic technologies do become tantamount to the markets they power.[14] In this respect, NASDAQ can be studied as a particular register of how electronic markets and technologies collide and collude in the making of electronic excess. The rhythm of electronic markets, as much as the processing speeds of microchips, impacts on electronic technologies' formation and transformation, distribution and erosion, both in terms of their materiality and value. The electronic, then, extends from technologies to markets and to modes of waste, decay, and disintegration. NASDAQ encompasses performative registers that are bound up with distributions and dispersals of matter. Using the term *electronic* to refer to markets describes not the absolute elimination of material resources but, rather, the mobilization and even more rapid turnover of materials and material relationships. These are electronic modes of waste, and this is how waste performs electronically.

Through these material and performative registers of electronic markets, there emerge distinct temporalities of exchange. The electronic exchanges that take place at the interfaces of NASDAQ terminals are typically urgent yet ephemeral. These modes of display, together with the interconnected network of exchanges, establish a pulse and performance that rework the formation of values. It is a network that arguably has contributed to the transformation of what value—or a commodity, particularly an informational commodity—even is. To assemble this natural history, I begin the next section with a discussion of several overlapping and ostensibly dematerialized screen displays and networks associated with NASDAQ. These displays span from the megalithic NASDAQ MarketSite building in Times Square; to the seemingly virtual and fleeting surface of the innumerable distributed screens where market transactions occur; to the networks, software, and automated technologies that inform this particular vehicle of electronic exchange. The screens, networks, and software that constitute NASDAQ emerge as material and performative infrastructures that impact on the rise and fall of electronic markets, the performance of electronic technologies, and the formation of electronic waste.

From Microchip to Megalith

The tale of dematerialization is often told through the rapidly shrinking size of digital technologies. Laptops now have more processing power than the computers that put astronauts on the moon and computers have diminished from room-size mainframes to compact and portable gadgets. But on the whole, the decrease in computing size has not, by any available evidence, reduced the total amount of resources deployed in the manufacture and consumption of digital technologies. Even though these technologies are smaller, they are consumed more frequently and in greater proportions.[15] So, by another process of "removal-transplant," the physical bulk from individual machines has diminished but has at the same time proliferated across more devices.

A similar transfer process seems to have occurred in the NASDAQ MarketSite headquarters in Times Square in New York, a location established to consolidate and present a "face" for what is otherwise a relatively decentralized and pervasive electronic market. In many ways, MarketSite is designed to reveal the sprigs and sprockets that make its engines turn. The designers selected for the project sought to convey a futuristic vision of NASDAQ and, to this end, settled on a design that would give MarketSite visitors a sense of inhabiting a computer. The designers note, "The client wanted a space that would look as different as possible from the paper-strewn New York Stock Exchange—one that resembles the inside of a computer." The design of MarketSite inverts the usual spatial relationship by placing people inside an environment that emulates a set of digital effects. Lighting within the spaces appears as "information traveling through a network" and is "strung on cables like microchips on a circuit board." Punctuating the ensemble, the design and lighting directs visitors toward an even more stunning feature. As a reviewer in *Architectural Record* describes,

> These lights are programmed to dim in a wave that draws attention first to a neon-lit, shimmering artwork of silk and metal fabric and then leads the eye through the space, which terminates at a curved 55-by-11-foot video wall comprising 100 video monitors. Continuously updated news, stock prices, and performance information are displayed at this state-of-the-art digital information system.[16]

The electronic stock exchange amasses as an architectural exclamation point, a concentrated and serial repetition of all the terminals that comprise its otherwise dispersed infrastructure. Inside this designed and materially recast network, it is possible to venture into an enlarged version of computers and circuitry and to experience a performance of electronic exchanges at the interface.

MarketSite was designed to be at once both "a physical environment that would help communicate the image of a company that has billed itself as 'the stock market for the next 100 years'"[17] and an "epicenter for financial and business news."[18] In order to convey the significance of this electronic market-without-a-market, however, a tremendous amount of material was deployed. The NASDAQ MarketSite tower is clad in what is declared to be the largest stationary video screen in the world. This surface, which is over seven stories in height, covers a span of nearly 10,000 square feet and is powered by nearly 19 million light-emitting diodes. The video screen displays advertising and NASDAQ messages and runs the ever-present virtual ticker tape of financial data across its surface. In what would seem to be a strange reversal to the dematerialization trend, microchips and computers have inflated to scales well beyond even that of the most prehistoric mainframes, into computers the size of skyscrapers, pixels at the scale of billboards, and data that is not virtual or immaterial but, rather, something we inhabit. Material structures shift not once but several times over. While NASDAQ is a dematerialized marketplace that conducts its transactions not on the trading floor but, instead, dispersed across telecommunication networks, it simultaneously inflates the electronic apparatuses of microchips, networks, and screens and rematerializes them at an epic scale. Through this inversion, NASDAQ appears to be "virtual" within an extensively material presentation.

As it turns out, an enormous amount of material and resources are required in order to establish and convey the sense of the virtual. The number of screens alone at MarketSite illuminates this paradox. From the hundreds of interfaces that spill over with the urgency of new economy news, to the roving electronic ticker tape that wraps the MarketSite building, to the millions of terminals worldwide that process and receive NASDAQ data, there exists a considerable concatenation of surfaces through and across which NASDAQ trading transpires. Although these are material architectures and technologies, they operate in support of the dematerialized imaginaries of electronic networks. From manufacture to display, the matter and the material operations of these screens are impalpable. These same screens eventually end up in the trash heap or are shipped near and far for recycling, but before they reach their final installments, they perform as the seemingly immaterial conduits for global finance.

Screening the Virtual

As an interface and space of transaction, the screen seems particularly conducive to conveying a sense of virtuality and dematerialization. Electronic objects collapse and disappear into the space and function of the interface.[19] Through the electronic transaction, the screen's role as a primary site of involvement seems to disappear from view, as the screen becomes a portal for a more virtual engagement. Yet as the array of screens, interfaces, and transactions at NASDAQ's MarketSite illustrates, the virtual is far from immaterial. The virtual, in fact, is a mechanism of expenditure. Such expenditure occurs most intensely in the apparently absent space of the screen, in what is the space of exchange.

Before I move further into describing the electronic infrastructure of NASDAQ, I would like to elaborate on the notion of expenditure, as it underscores the key ideas in this chapter. The "virtual" of course has a long history of use, from the potential or germ of possibility to the more general sense of a simulated reality as is typically meant in the context of computing.[20] The virtual also at times has come to mean an abstract model or paradigm to which practice is made to conform.[21] Without plunging into the vagaries of these uses and attempting to resolve the virtual, I would like to make a lateral move and suggest that the virtual as it emerges in this specific discussion of NASDAQ refers neither to the material nor to the immaterial exclusively, neither to model nor to practice specifically, neither to potentiality nor to actuality, but, rather, to expenditure. While the virtual appears to exist in a "space of flows," generally unfettered and detached from material structures, the expenditure that the virtual enables has consequences that exceed the material or immaterial (and, as such, undoes this division). The sense of the virtual that emerges in the allure of NASDAQ's MarketSite is the expenditure required to sustain and circulate the forwardness of digital technologies. Often, in the "forced march" of technological innovation and growth, more is expended than is gained (as will be discussed shortly concerning financial outlays for digital networks and technologies). The virtual is the force of expenditure that is ultimately required to sustain the momentum of technology and the momentum of its promise (because the two are inseparable).[22]

This expenditure is seemingly abstract, but it in fact constitutes an intensity and performative force through and around which electronic markets realign.[23] In some respects, expenditure can be seen to be a defining trait of the "new economy." As much as $150 billion was raised during the mid- to late-1990s to support and galvanize the new economy.[24] Credit, speculative or venture capital, and stock offerings are examples of the continuity between expenditure and the virtual. These forms of finance impact on the movement and amplification of markets and market activity. They are neither fact nor fiction; rather, they are virtual expenditures that set in motion self-perpetuating and even obligatory economic conditions.[25]

The enormous sums of money moving through markets and into technology companies and the ensuing "speculative bubble" that resulted in an overinflated NASDAQ came to be known, after Alan Greenspan's characterization, as "irrational exuberance." With the collapse and correction of the "new economy," it became difficult to verify the extent to which new technologies and the new economy created conditions of demonstrable economic growth. Economist Robert Shiller suggests that whether there is measurable growth stemming from the new economy is perhaps less important than "the *public impressions* that the revolution creates."[26] Through repeated online activity or through the presence of multiple electronic interfaces scrolling financial news, a self-reinforcing logic may emerge that can be located neither in the impressions nor in new technologies but, rather, in the expenditure (in time and money) required to keep both of these afloat. Digital technology is meant to constitute a "new growth paradigm," and this objective may become the guiding agenda through which electronics and electronic exchanges operate.

Screens are a site of intensive practice and attention through which growth-focused electronic exchanges transpire. Expenditure at the interface is not just restricted to an excess of financial outlay in the rapid exchange of shares through electronic markets, however, but also receives yet another source of reinforcement from the reporting of financial news. From CNN to CNBC, the media screens of financial news intersect with the electronic screens of market exchanges, at times even collapsing into the same space, as brokers watch financial news while trading.[27] At this juncture of media screens on digital screens, it is essential to recall that one of the primary functions of NASDAQ's MarketSite is to serve as a media site, a space where "major financial broadcast outlets conduct daily reports from MarketSite and reach viewers around the country and world."[28] The number of these "live market updates," typically broadcast by major media conglomerates, reaches over 175 per day. So pervasive and insistent are these broadcasts that they come to seem as essential to the new economy as the technology and markets on which they report.[29] The speed and prevalence of the electronic ticker tape and the insistence of media screens contributed to the reordering of finance and its performance.[30] The financial news media are not only entangled in the "irrational exuberance" of the new economy; they also help to generate the terms of the new economy's performance.

While in the 1990s these screen-based performances of the new economy may have been relatively novel if not futuristic, they are now increasingly distributed across multiple spaces where the business traveler may be in transit. Media screens laden with financial information, whether in the form of scrolling indices or news analyses, distribute across a wide landscape that includes, as geographers Gordon Clark and Nigel Thrift identify, "hotel chains around the world, airport lounges, and shopping malls," as well as "laptops, PDAs, and mobile phones," which allow for updates on investments and financial news "on the move."[31] These media screens have become constant indicators of the status of markets. They have fueled the performances of expenditure (and exuberance) that variously circulate as new economy speculations. From these overlapping infrastructures, networks, and technologies, there emerges a mode of electronic exchange that is so pervasive it seems to fade into a flickering background noise. Part of the reason for this persistent hum is not just the sheer everydayness and everywhereness of these networks but also the rapid and fleeting pace at which they operate.

As an electronic stock market, one of NASDAQ's primary distinguishing functions is its unmatched speed of exchange.[32] The market's trading networks are fast and comprehensive, linking traders in 146 countries. To improve their "transaction services," which are the "engine" of their market, NASDAQ acquired an additional electronic communication network (ECN) in 2004, which further improved its efficiency and increased its liquidity. In 2007, NASDAQ averaged 2.17 billion trades daily. The NASDAQ systems are capable of processing 250,000 messages per second, an average of 1 millisecond each.[33] Described in these estimates is a pace of exchange that is bound up with a capacity for rapid rates of circulation, where shifts in value tick across screens and terminals with an ephemeral and fleeting insistence.

Electronic markets can thus be characterized by higher rates of stock turnover and increased volumes of trading. These accelerated levels of electronic market activity can be traced to an increase in online trading in general, as well as to greater accessibility and ease of making trades, together with more constant news about financial activity.[34] These assessments add up to a certain rhythm of economic life.[35] Economic progress becomes defined through rates of transfer. Electronic markets facilitate more rapid rates of transfer, but in so doing, they alter the materiality and performance of those markets. Just as electronic networks enable trading in greater speeds and larger volumes, so this increase in speed and quantity potentially results in greater volatility. But it is precisely through the sudden and even minute shift in values that profit may be made.

With the migration of trading from the physical floor of a stock exchange to electronic networks, the ups and downs of market values are tracked within different scales and temporalities. While traders in an open pit depend on a commanding physical performance in order to execute trades, electronic markets engender a much different attention to and manipulation of trades.[36] The ephemeral shifts in electronically displayed values can translate into money lost or gained. Anthropologist Caitlin Zaloom describes, through comparative ethnographic research, just how intently traders play the spread between bid and ask prices by continually negotiating "temporary assessments of market conditions, momentary markers of approximate valuation."[37] The speed of trading becomes bound up with the rates of transfer and tracking afforded by electronic technologies. What traders must accustom themselves to most of all is the instability of these values. So while they work within instability, they also turn it to their advantage. Electronic technology, which amplifies instability in many ways, also becomes a way to take advantage of the ambiguities and volatility of numbers.[38] The question is whether the traders are playing the numbers, the technology, or both (or whether, even, the technology is playing them). The rapid scroll of financial data across screens can be tracked, momentarily stabilized, and acted on through these electronic devices. While the values operated on may seem relatively fleeting, this process is a material performance, involving electronic screens and networks, traders' bodies, and office buildings, distinctly electronic temporalities and rhythms of exchange.

In this discussion of the volatility and volume actualized by electronic markets, what we take for the virtual—for apparently dematerialized conditions and objects of exchange—is in fact closely bound up with material and temporal expenditure. The ephemerality of numbers on which profits are won or lost and the errant spikes and dives in value emerge from and contribute to a sense of dematerialization and destabilization. This sense of time, of volatile and instantaneous events continually renewed, resonates with what Haraway calls the "technopresent" where "beginnings and endings implode."[39] This is a temporality that describes a rate of turnover, a rhythm of exchange, and an anticipation of progress that could be described as coded and so flattened, characterized by a sort of automaticity. Increased speed and expenditure give rise to the sense of dematerialization that is so specific to electronic technologies and electronic markets. The technopresent describes time as a program, which is operational and efficient but also dematerialized and ultimately depoliticized.

Dematerialization: Networks and Software

While it is by now clear that dematerialization is in fact a contradictory way to describe electronic technologies that are in fact deeply material, there are of course clear reasons why these technologies do seem to dissolve. From thin screens to tiny chips and from dispersed networks to rapid rates of exchange, many of the qualities of electronics convince us that they are relatively free from material requirements. Yet the term *dematerialized* does not necessarily mean "without material" but may, instead, refer to modes of materialization that render infrastructures imperceptible or ephemeral. This is electronic technology's sleight of hand, its magic. It appears to be immaterial, but this sense relies on dispersed material infrastructures. Such a condition does not simply involve revealing the invisible but obviously physical props that enable these apparently virtual technologies. Instead, a sense of immateriality is bound up with complex and specific ways of mobilizing and imagining material performativity as being free from resource requirements.[40]

Dematerialization can further constitute a way of making technologies seem even more operational and effective.[41] The sense of dematerialization, in this case, may emerge through the speed of exchange and space of the interface, which foreground the transfer of signals and light in place of the supports of chemicals, metals, plastic, and labor. Here is a process of erasure—as well as a process of substitution that works toward a new performativity in the form of accelerated exchange and output. Such erasure unfolds through the speed of electronic networks but also through the apparent immateriality of the software that influences the "functionality" of those networks. Yet another form of erasure occurs in the timing of these exchanges, as suggested earlier. The ephemerality and accelerated rates of exchange that electronic networks facilitate influence, in turn, how we understand the materiality or immateriality of digital technologies.

Rather than refer to dematerialization, in relation to markets Don Slater suggests that we consider how things hold together at all. He instead proposes that objects and goods may move through processes of "stabilization" and "destabilization."[42] The market is a primary space where this operation takes place; it is an institutional and authoritative register for informing the stability of goods and lapses in value. Dematerialization describes less a condition of things without materiality, in this sense, and more the processes of materialization that allow things to register as entities. How and why do objects hold together, and what resources are at play in both stabilizing and destabilizing those objects? Beyond the dubious category of "physical" objects, what other dynamics emerge to reveal how "things" like computers, mobile devices, software, microchips, screens, NASDAQ indices, and billboards register as sites of momentary value and materiality—or immateriality? Electronics may even appear to be dematerialized because they are more fleeting, more disposable, "provisional," and even volatile.[43] Provisionality, ephemerality, and volatility have arguably become more central qualities of goods and markets; these are qualities that may contribute to a sense of dematerialization, and they are also mechanisms for realizing a perceived increase in performativity within the new economy.

Even prior to the establishment of the NASDAQ network, financial institutions were some of the first organizations to employ computerized and automated telecommunication networks in order to facilitate the processing and automating of transactions. Nearly parallel to these usages, manufacturing companies began to take up the use of these networks in order to ensure more regular control of stock and inventory.[44] These histories will be taken up in greater detail shortly, but this discussion of networks begins with the most dematerialized version of networks—as they are imagined to be in an indefinite but dematerialized future. Kevin Kelly, one of the founding editors of *Wired* magazine, suggests that networks allow not just for the more effective coordination of manufacturing but also for the potentially complete dematerialization of systems required to produce things in the first place. In Kelly's assessment, goods may be developed according to "'just-in-time' production techniques," which could "respond to trends in consumption."[45] But in order for such timing and responsiveness of production techniques to be enacted, networks must be employed. Networks allow for sudden changes, responses, and adaptations that can be set in cue with market demand. To realize such responsiveness, however, these networks must become not only quicker but lighter. Kelly writes,

> But this flexibility demands tiptoe agility from multi-ton machines that are presently bolted to the floor. To get them to dance requires

> substituting a lot of mass with a lot of networked intelligence. Flexibility has to sink deep into the system to make flexible manufacturing work. The machine tools must themselves be adjustable, the schedules of material delivery must turn on a dime, the labor force must coordinate as a unit, the suppliers of packaging must be fluid, the trucking lines must be adaptable, the marketing must be in sync. That's all done with networks.[46]

As much an advocate for as an analyst of dematerialization, Kelly sets the tone for a more immaterial economy by promoting the advantages and efficiency of these seemingly lighter networks. Automation here occurs through dispersed networks, which makes objects cheaper to manufacture and reproduce, because manufacturing is faster, the objects may be smaller, and the processes require less material. So promising is this ostensible elimination of material inputs that Kelly forecasts a time when "one can imagine the future shape of companies by stretching them until they are pure network."[47] As pure network, companies would be pure process, and any material they produce would always be in transition, transformation, and exchange.

Yet, for all their seeming absence of material requirements, networks have been major sites of resource expenditure.[48] So convincing is the logic of networks for their ability to improve efficiency, capacity, timing, and profits that scores of companies have invested in network technology in pursuit of this promise. Don Schiller documents how in the 1990s, at great cost, a number of companies undertook network application projects in order to save time and money and to speed products to market.[49] The majority of companies investing in these technologies have been located in the United States, where expenditure on information and communication technologies soon surpassed that of any other capital expenditure. Despite this investment in network and information technologies, many of these ventures often did not achieve their stated aims. Far from constituting a reasonable investment or restructuring of production and distribution, these network projects were then characterized by tremendous expenditures and waste. Many of these network projects were in fact never completed.[50] A tremendous amount of money and resources was expended in order to implement the logic and technology of networks. Such expenditure, even when it fails, appears to be a way to reinforce the promise and prevalence of electronic networks. This expenditure has had such an impact, moreover, that, together, these information technologies have now been classified as the largest industry in the United States.[51] With such a sudden and thorough rise to a dominant position, information and network technologies have contributed to the transformation of economic practices and manufacturing conditions alike.[52]

If networks describe the restructuring of economic, material, temporal, and environmental processes, then how do we begin to describe the qualities of such restructuring? These networks enable a sense of virtuality, of greater efficiency, accelerated speeds, and lower resource requirements.[53] At the same time, networks emerge not as materials or resources but as relations and systems of exchange. Even though networks have even been referred to as the new factories, as a "factory for information," the prevailing sense is that a network somehow describes modes of operation rather than sites or materialities. Yet the tendency toward apparent dematerialization is a key part of how a network operates. Kelly elaborates on the network-as-factory theme: "A factory-made widget once followed a linear path from design to manufacturing and delivery. Now the biography of a flexibly processed widget becomes a net, distributed over many departments in many places simultaneously, and spilling out beyond the factory, so that it is difficult to say what happens first or where it happens."[54] While resource inputs and the space of manufacture become decentralized through a network, Kelly's statement suggests that a widget is not without resource requirements but that those resources have been distributed in different ways, across networks.

In this sense, materialities are restructured in a way that changes their ratio and distribution, as well as their economic, political, and environmental effects. A network may redistribute material, but it does not eliminate it. A network still requires resources, and it is essential to take into account the resources it extracts, processes, and distributes and where the wastes from processing circulate. The "network" of electronics extends from Superfund sites in Silicon Valley, to the networks of exchange and valuation of NASDAQ, to the recycling villages and dumps in China and Africa. It is through these other expanded networks that it is possible to trace out these transformed material structures and these electronic modes of waste. These networks not only are made of more than sunshine and signals; they also depend on hidden labor, political inequalities, and environmental damage. But the distribution of these aspects of electronic networks can be disparate and remote from sites like electronic markets. Electronic markets, moreover, typically operate through programs of efficiency—or software—that can render automatic and even seemingly "natural" many of the functions, distributions, and relationships that make these exchanges possible.

On the surface of things, NASDAQ may exist as an electronic network, but in order to actually access it, users require distinct software that will allow them to access discrete "levels" and modes of market information. Software exists for "data feed" and for "transaction services." In fact, it is software that enables the operations of computer networks, by programming for specific capacities and "functionalities," including algorithmic trading.[55] Software enables another level of material inversion, not those megaliths constructed to resemble microchips, but seemingly immaterial architectures constructed to power vast material and manufacturing structures. Software is the code that appears to circumscribe the ratios and proportions, the speeds and relationships, within networks. The critical function of software is to program processes—of manufacture, calculation, automation.[56] What drives networks is software; this is the automatic program that constitutes the design of the manufacturing process. In fact, most expenditure is now directed toward things that look increasingly like software, from research to licensing, but these inputs typically do not fully register within economic processes. Invisible though it may seem, software still operates in the microspaces, networks, and unnoticeable backgrounds; in the "guts of a set of commodities"; and, finally, across multiple platforms to be delivered as programmed content to screens everywhere.[57]

Software ensures that the lid stays on the black box of electronics, and our only window into these mysterious devices is through the interface, which can effectively obscure the workings of this technology. This directing of attention toward the effectivity and functionality of these devices and not toward their resources, labor, and environmental effects is a way in which software programs matter. But in programming matter, software becomes tied to matter; it constitutes a distinct articulation of material processes. In this respect, it may even make sense to say that "there is no software."[58] There is no software because there is nothing soft—or absent—about it. Media theorist Friedrich Kittler explains how the difficulty of determining just what software could be has even led to its near extinction in German regulatory spaces, where the "concept of software as mental property" had to be rescinded, as it was next to impossible to determine where hardware stopped and software started, since the latter could never operate "without the correspondent electrical charges in silicon circuitry."[59] As soon as we attempt to delineate software, it inevitably leaks into material structures, demonstrating that while the program of software operates as a code of effectivity, it is irrevocably bound up with material and technological processes that enable these performances. Software facilitates the increasingly refined programming of matter and exchanges; but even more, it allows for the sense of expanded possibilities for transforming that matter—to dispense with it, to distribute it, and to generally minimize material requirements so that the process itself can appear infinite, even if the resources are not. The "program" of automation may help to explain further why this tension between material structures and apparent erasure has such a lasting influence on the performativity of electronic networks.

Automation: Programming Matter

The soft and hard technologies that fuel electronic markets were a long time in the making, and depending on which influences we would choose as most critical, we could find major contributing factors in the nineteenth century, with Charles Babbage and his Difference Engine, or in developments during the World War II era, including the Turing machine and ENIAC. But the advent of the second wave of automation, in the 1950s, may most directly inform this analysis of electronic markets. Automation allowed for the control of stock and inventory and began the movement toward "automatic programming" that would enable machines to coordinate entire financial and industrial processes without human intervention. Taking up the term *automation* and applying it to manufacturing and businesses alike, John Diebold used the notion as a tool for rethinking economic processes through computerized feedback. Of this new model of 1950s industrial practice, he wrote, "The push-button age is already obsolete; the buttons now push themselves."[60] Automation is relevant to this investigation into electronics not just because the first mainframe to be applied to industry and financial use, the UNIVAC, was employed by General Electric and NASDAQ alike but also because it was within the theories of automation that notions pertaining to programmed exchange and a dematerialized stock exchange were first developed.

In the same book in which he popularized the term *automation,* Diebold put forward a proposal to rethink "the problem of the New York Stock Exchange" through automation.[61] The NYSE required more than just the mere appendage of some "new gadgets" to what were "obsolete processes," he argued; instead, the stock exchange needed to rethink its entire operations through automation. Diebold elaborated on what he perceived to be the inefficient and outmoded operations of the NYSE.

> Characterized as the nerve center of American industry, the exchange is really a glaring anachronism. On the floor of the exchange as in the ancient market places, the traders stand at their posts and offer wares—not stone jugs, but stocks and bonds. Hundreds of men swarm over the paper-strewn floor. Messengers dart to and fro with scribbled bits of paper. The glitter of a few modern devices such as the high-speed ticker tape (which records what has happened but does not participate in the action) is so blinding that we never question the basic process.[62]

Diebold sought a way in which to "automatize" the materially encumbered exchange. He suggested, "What is called for is something completely different from the exchange floor as it exists today." That something different was the use of computers to execute automatically the processes of exchange. Such a change would be so revolutionary that computers might even "provide a means for eliminating the exchange floor altogether."[63] Diebold suggested that automation would greatly improve the speed and efficiency of the stock exchange. As part of this improved operation, the required material infrastructures would be expendable and even eliminated. In Diebold's description of automation is the logic that later comes to define the workings of software and networks, and of electronic market transactions.

In Diebold's text, where he searches for early applications for automation, it is the elimination of existing material structures and relocation of processes through programmed machines—in other words, computers—that would allow for the realization of greater efficiency, not just in the circuits of exchange, but also in processes of manufacturing. Computers were seen not only as a way to improve speed and efficiency through automation but also as a way to reduce waste and free workers from repetitive tasks.[64] Elaborating on these advantages, Diebold suggested that automation involves more than simply making existing products through computerized means. Instead, automation would lead to automatic processes that would, in turn, inevitably change the products produced.[65] These alterations are due not just to automation but also to the electronic quality of the machines doing the processing.

Electrical and electronic automation can lead to entirely different "inventories," comprised, as McLuhan suggests, "not so much of goods in storage as of materials in continuous process of transformation at spatially removed sites."[66] The removal, redistribution, and transformation of goods through these processes apply not just to automation in the 1950s and 1960s but, arguably, just as well to electronic networks of finance and industry in operation today. Just as with Kelly's notion of a "pure network," when materials are in constant transformation, they seem to dissipate completely. But if we look closely, we see that the materials have not just disappeared; they have instead realigned and transformed—stabilized and destabilized—through electronic modes of exchange.

Automation, from industrial-mechanical to information-electronic, is a process that transforms matter—it could even be called, following philosopher Michel Serres, "a revolution operating on matter."[67] When technologies are automatic and autonomous, they become catalysts not only of material complexity but of new distributions and creations of energy.[68] Electronics and electronic networks—coded, distributed, efficient, automatic, and seemingly immaterial—give rise to distinct patterns of movement, exchange, and transformation. When machine technologies spark conditions of material transformation and complexification, they seem to operate as "natural" forces. This is exactly the sense in which I here deliberately take up the term *natural* to write toward a natural history that describes processes of materialization as situated, cultural, political, and environmental events. This materiality describes not an essential or given condition but, instead, a technonatural enfolding, where electronics generate distinct material processes.

Exchange Theory

Exchange, the processing that transpires across electronic networks, becomes the basis not just for transmission and transformation but also for deformation. Serres writes that "the exchanger is also a transformer," and so the process of exchanging messages becomes a process of change.[69] Within electronic markets, transformation takes place in particular ways: toward the instantaneous, the voluminous, and the volatile. Transformation and expenditure may give rise to destabilization. Yet within electronic markets, this processing and circulation becomes the basis for value. Instability and volatility can actually become forces on which to capitalize. Exchange, in this sense, can be understood as the source of value. The ways in which objects circulate—or are exchanged—inform their value.[70] By focusing on exchange, we can study not just how commodities form but also how they circulate in and out of value. Such an approach allows us to go beyond the object or product and, instead, to consider how exchange can enable objects to obtain value as commodities and, by extension, how exchange can also ensure the loss of value and potential decommodification of objects.[71] Indeed, in the context of this chapter that focuses on the sorts of exchanges that electronic networks enable, it becomes evident that the terms of exchange, value, and commodities shift. The processes of networks and software direct attention toward process as a key register of products. The rates and the volatility or provisional quality of exchanges can also enable more rapid processes of valuation and devaluation. In the language of the new economy, the commodity may no longer even be a stable object but may instead be formed through a networked process.[72] Within electronic processes, commodities have become marked by instability, a certain alchemy, which accelerates the process of transformation, information, and deformation around the boundaries and values of those goods.[73]

Some of the earliest attempts to theorize just what information—or an information commodity—is and how we should measure it for its economic value have focused on the fluid, rather than solid, aspects of information. Indeed, Fritz Machlup, an Austrian-American economist who contributed to the popularity of the phrase *information society,* asked in 1980 whether there were "any ways to measure or estimate the magnitudes of the stocks and flows of knowledge."[74] Machlup was concerned with how to establish a common standard of measure for anything that could count as information, which at that time meant "society's stock of recorded knowledge, mostly in the form of books and journals stored on the shelves of libraries." Unlike the "How Much Information" report discussed in chapter 1, Machlup found this physical basis for measurement to be insufficient, because "counts of volumes and counts of titles lead to very different results."[75] Knowledge counts could be easily duplicated, and the scale at which these knowledge counts should even begin was not obvious: should we count works, pages, titles, or volumes? Information challenges the traditional units of measure, which in this case were strictly tied to physical formats; instead, those instruments have to be invented, modified, and adapted to the task of measuring an apparently formless entity that does not compare to the regularity of stock.

Wrestling with this problem, Machlup decided that "flow" is the most ideal measure for reckoning with the quantity of "society's knowledge." By measuring circulation, it is possible to measure value.[76] The measure of value adheres not to the actual unit of information but, rather, to its circulation; its circulation implies exchange, and exchange equates to value. If information is requested, transmitted, or received, then it is in use or in demand, and it therefore moves within structures of value. These structures of value are arranged as networks. This is how a network can further enable value by increasing the web of connections. As Kelly writes,

> If you have the only fax machine in the world it is worth nothing. But for every other fax installed in the world, your fax machine increases in value. In fact, the more faxes in the world, the more valuable everybody's fax becomes. This is the logic of the Net, also known as the law of increasing returns. It goes contrary to classical economic theories of wealth based on equilibratory tradeoff. These state that you can't get something from nothing. The truth is, you can. . . . In network economics, more brings more.[77]

As Kelly describes, circulation—in the form of networks—is not just the means for generating value; it is the source of accumulating value. But such structures of circulation, accumulation, and value do not describe information alone. Even noise—junk messages—can acquire value through circulation.

As I have suggested early on in this chapter, circulation is the basis not just for value but also for devaluation; as such, it is bound to the generative dynamic of waste. What appears to be waste may even acquire value through its circulation within particular spaces of value. This condition is true for both spam and junk mail, which constitute a large proportion of Internet and mail traffic. While attempts have been made to legislate against the circulation of junk, estimates still refer to nothing less than an exponential increase in spam, or unsolicited e-mails. Billions of spam messages circulate through the Internet, a volume that is made possible by innumerable personal computers that are programmed to inundate the electronic networks of the Internet. Spam is a program as much as a sham offer for property in Bermuda; it automates the circulation of messages in bulk across networks that do not—up until recently—discriminate from the information or noise that it exchanges. Spam is lucrative precisely because it flows in massive quantities. By sheer odds, some messages do eventually reach receptive audiences, who execute "buy/

sell" orders (most likely based on "pump and dump" missives).[78] What informs the circulation of these messages most of all is the fact that they are part of an automated exchange made in bulk, where the volume of material in circulation eventually realizes a profit by finding its way to spaces of value and exchange.[79] Just as the volume and frequency of these exchanges may actualize a profit, however, so may they circulate through spaces of devaluation. In just this way was it once possible, with the dot-com crash, for NASDAQ to be valued as nothing but junk.

In an even more pronounced performance of these cycles of valuation and devaluation, the financial crisis that has played out since the end of the 2007 housing bubble, fueled by subprime mortgages, and through the ensuing credit crisis has generated its own cast of material remainders. From collapses in balance sheets to mortgage foreclosures and loss of jobs, multiple spaces of devaluation have unfolded within the mysterious calculus of speculative capital. Complex financial instruments and distributed investment packages have, in many ways, been amplified through the infrastructures of electronic markets and exchanges. The scale of the current market "correction," with write-downs and write-offs in the trillions of dollars, has a discomfiting correlative in the now-vacant homes, closed storefronts, unemployment lines, and idle container ships that scatter from the swamps of Florida to the harbors of Singapore.[80]

To understand the fallout from the rise and fall in value, from so many numbers flickering across screens and processors, it is necessary to understand what role waste and wasting play in this dynamic. Waste operates not just at the terminal end of a commodity's life but across its production, exchange, and consumption.[81] When mapped through these more extended processes, exchange emerges in a more entangled relation with waste, both in the ways devaluation occurs and in where the potential for revaluation resides. This is a way of reading exchange through the dynamic potential of waste. As cultural theorist John Frow elaborates through his reading of Thompson's *Rubbish Theory*, "the transformation of value is not grounded in the intrinsic properties of objects"; rather, value emerges as "an effect of the circulation of objects *between* regimes of value."[82] These circulations are complex, possibly driven as much by wastefulness as by the recuperation of value. But such circulation cannot be reduced to markets alone, because the emergence of value through circulation works within spaces of potential virtual expenditure. Virtuality is bound up with the inexhaustibility of things and with the generative and dynamic qualities of waste and the formation of value.[83]

Waste is at once an inevitable and distinct force at play, informing the circulation of objects and their value. Waste overlaps with other circuits of exchange, other networks and material distributions. In this sense, it is not too far to trace another connection between the circuits of electronic exchange to the resurfacing of electronic waste as it circulates toward another exchange, the circuits of disposal and recycling.

Dematerializing and Rematerializing

The circulation of waste extends from the "virtual" and performative exchanges of electronic markets to the material and environmental exchanges of digital rubbish. The apparent dematerialization of digital technologies may enable greater "functionalities," but in many ways, it also generates greater volumes of waste. As the seemingly more immaterial digital technologies demonstrate, this is due, on one level, to an enhanced ability to process and distribute materials.[84] By some odd turn of events, processes of dematerialization have even facilitated accelerated rates of output.[85] As this chapter attempts to establish, however, these same processes that seem to require less resource-intensive production and exchange rematerialize not just through abundance, toxicity, speed, destabilization, or performativity of materials. Electronics rematerialize again through obsolete devices in the form of electronic waste. Indeed, electronic waste gives rise to a reconsideration of what constitutes the boundaries of electronic technologies, which intersect with processes of materialization from exchange to automation.

To "rematerialize" electronic technologies is also to map the political relations that support their operations. The politics of dematerialization emerge in sharper focus when we consider where the overlooked remainders of electronic technologies circulate. As mentioned earlier, much of the electronic waste that is sent for "recycling" from the United States and other wealthy countries finds its way to less economically privileged countries. The flow of garbage typically follows this course from developed to developing country. This circulation and exchange, delineating the valued and the devalued, sustains the figure of dematerialization. The ability to sustain economic growth may even require the sense that growth has a more "immaterial" quality; yet supporting this immateriality is a politically unequal material infrastructure that enables growth.[86] To this extent, the Basel Action Network has suggested, in its report on the exportation of electronic waste to Southeast Asia, that much of the "virtuality" of digital technologies exists by virtue of the factories and dumping grounds that are positioned in locations remote from sites of consumption. By rematerializing electronic technologies, it is possible to draw together these apparently disparate relations as constitutive material processes.

Strategies of rematerialization can be one way to locate the apparently dematerialized flows of the digital. Just as the interface fades from view, a conduit for the exchange of so many electronic messages, it comes into focus once again in the form of inert and abandoned computer monitors and abandoned screens of all types. Many of these screens are composed partly of recyclable materials—glass and copper yokes. But the process of their extraction is toxic, and this extractive labor is typically performed not by users of computers or electronic screens but by workers who bear an entirely different relationship to these machines. In contrast to the relative disentanglement of computer users, these workers' "place of work," as media theorist Lisa Parks writes, "has become the inside of the machine—the part that is kept off-limits, locked up, closed off in Western consumer societies."[87] Beyond the interface, there are extended global economies through which discarded computers are processed. The labor, bodies, and economies bound up with dismantling computers entail a much different relation to the interface and to the black box of electronics. The workers who dismantle monitors typically extract the cathode-ray tube (CRT), a device rich in copper but also highly toxic to remove.[88] Images and exchanges that processed in milliseconds transform into metal scrap to be salvaged for raw materials markets. Far from constituting a virtual space, the apparently dematerialized interface depends, in fact, on power structures, resource movements, and material economies—all of which rematerialize when electronics literally break open and become waste.

Captured in this chapter are the sites and processes that are revealed by moving from the glow of the interface to the "inside of the machine" and beyond. From the initial discussion of inhabiting the megalithic microchips of NASDAQ's MarketSite to the screens, networks, and software that enable programs of automated exchange, electronic exchange relies on the displacement, dematerialization, and destabilization of technologies, as well as the generative dynamic of waste. The interface rematerializes as an electronic technology bound up with these performative registers—as well as with the global economies and ecologies of resource inputs and waste disposal. The material effects of discarded electronics often register far from the spaces of their past operation. "These sunshine-belt machines," as Haraway writes, "are as hard to see politically as materially."[89] When they are rematerialized—mapped within a layered natural history—they emerge as complex material and political devices. The next chapter turns to the circuits that enable the consumption and disposal of so many electronic interfaces. These infrastructures, which undergird and coexist with the more performative and distributed electronic exchanges discussed in this chapter, rematerialize electronics from networks of signals and light to the often extended and complex circuits of material divestment and disposal.

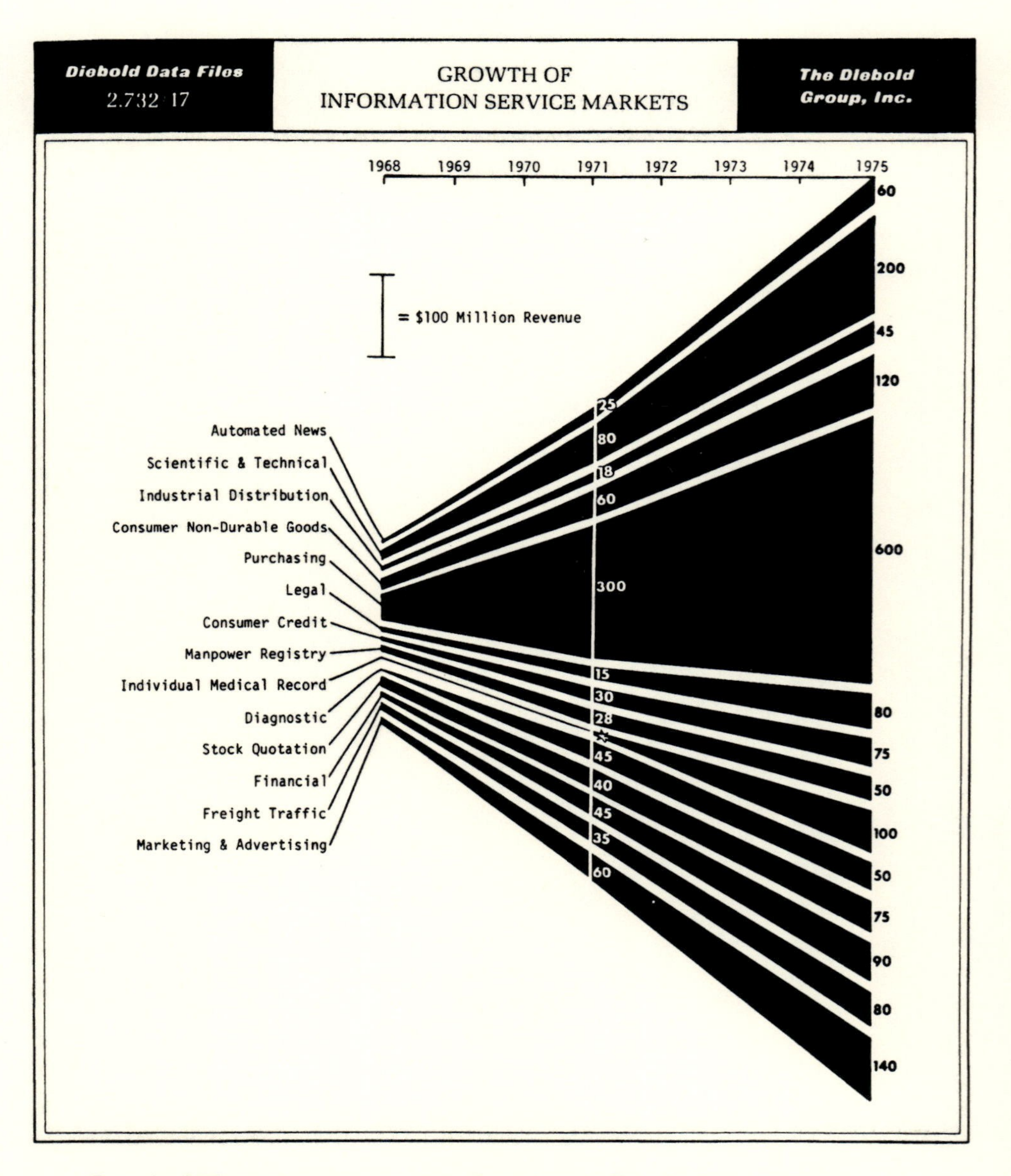

Growth of information service markets, from the 1967 Time Incorporated report "Information Utilities as a New Business Opportunity: Management Summary," Charles Babbage Institute, University of Minnesota. (Courtesy of the Diebold Group.)

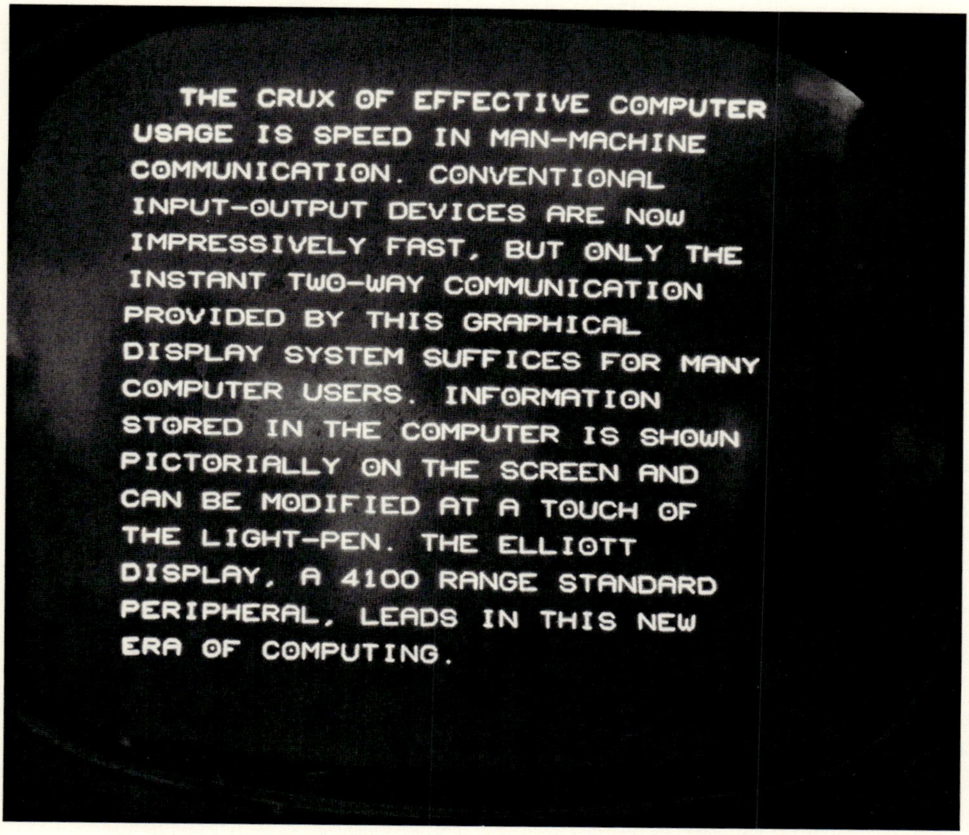

THE CRUX OF EFFECTIVE COMPUTER
USAGE IS SPEED IN MAN-MACHINE
COMMUNICATION. CONVENTIONAL
INPUT-OUTPUT DEVICES ARE NOW
IMPRESSIVELY FAST, BUT ONLY THE
INSTANT TWO-WAY COMMUNICATION
PROVIDED BY THIS GRAPHICAL
DISPLAY SYSTEM SUFFICES FOR MANY
COMPUTER USERS. INFORMATION
STORED IN THE COMPUTER IS SHOWN
PICTORIALLY ON THE SCREEN AND
CAN BE MODIFIED AT A TOUCH OF
THE LIGHT-PEN. THE ELLIOTT
DISPLAY, A 4100 RANGE STANDARD
PERIPHERAL, LEADS IN THIS NEW
ERA OF COMPUTING.

Elliott 4100 display monitor, ca. 1966, Science Museum of London. (Courtesy of Fujitsu.)

THREE

Shipping and Receiving

CIRCUITS OF DISPOSAL AND THE "SOCIAL DEATH" OF ELECTRONICS

Nothing good is endless in the computer world.
—J. DAVID BOLTER, *Turing's Man*

The "Social Death" of Electronics

Electronics eventually circulate toward other spaces of exchange that are situated far beyond those apparently dematerialized interfaces discussed in the last chapter. Electronic technologies that once powered markets reach obsolescence and are discarded. The outdated debris of computer monitors, printers, hard drives, power cords, peripheral storage devices, mobile phones, and servers that make up electronic networks eventually lingers in assorted stages of disposal, from the warehouse to the rubbish bin. Disposal is a continuation of the transmission and processing of electronics, albeit within distinctly different formats. Disposal is formative in the making and unmaking of the materiality of electronics. The practices of disposal involve multiple modes of material disassembly and depend on interconnected geographies for the circulation and recuperation of discarded devices.

Two narratives concerned with the practice of disposal indicate the potential scope of these material and geographic circuits and practices of disposal. In *Invisible Cities*, Italo Calvino describes a metropolis, Leonia, which refreshes itself by discarding all its objects on a daily basis. So persistent is the process of using up and expelling goods that this becomes Leonia's defining attribute, its apparent source of pleasure. The city's constant stream of refuse is transported by anonymous workers to unknown places located on the urban periphery. Yet the practice of

Electronic waste dismantling of monitor to remove copper yoke, Guangdong, China, 2002. (Photograph courtesy of Basel Action Network.)

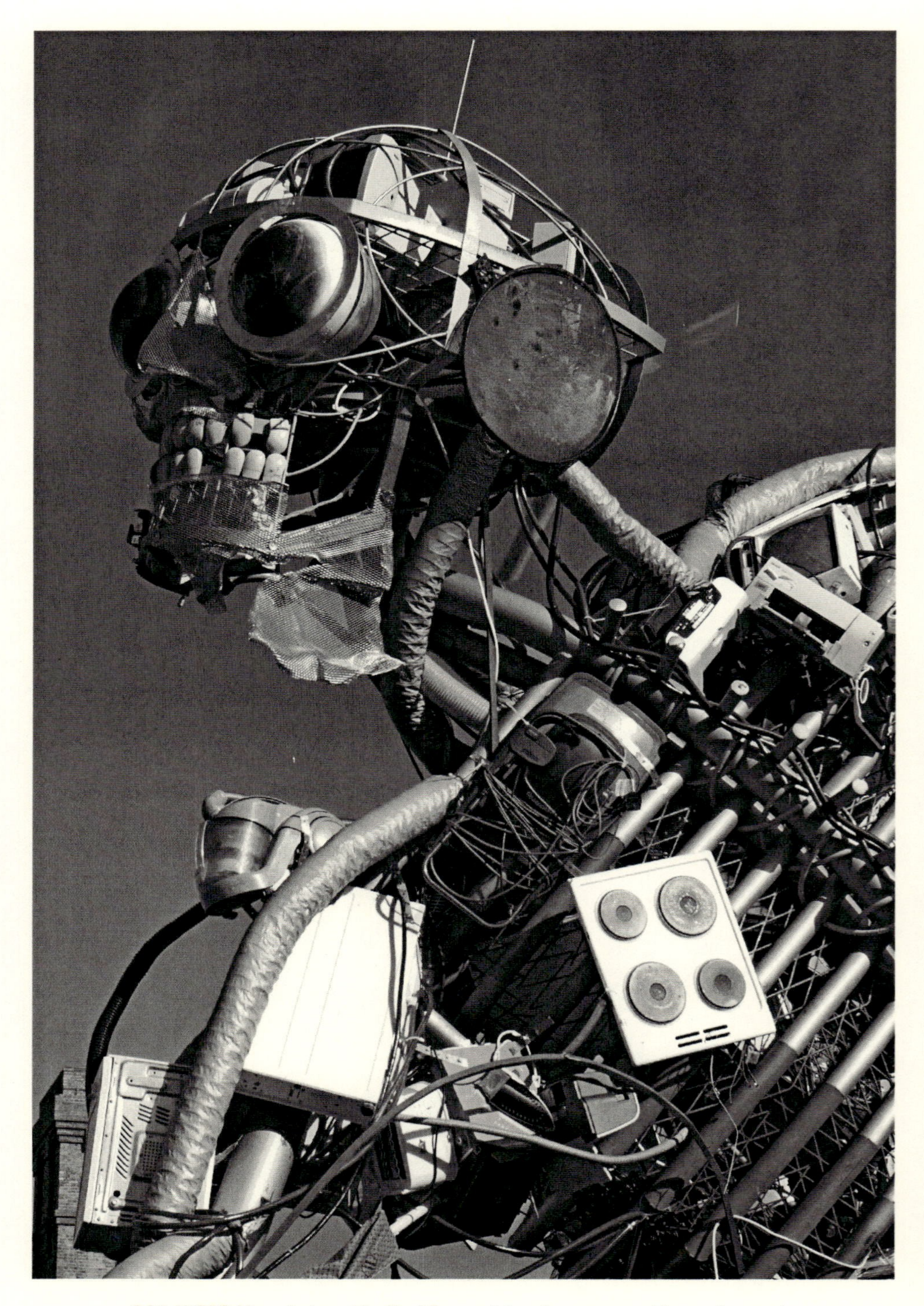

RSA *WEEE Man,* designed by Paul Bonomini and constructed from electronic appliances, London, 2005. (Photograph courtesy of the Royal Society for the Encouragement of Arts, Manufactures and Commerce / David Ramkalawon.)

expulsion grows to such epic proportions that an increasing quantity of debris accumulates and threatens to unleash in a cataclysmic landslide. In the process of disposing of its remains, Leonia unwittingly constructs orders and spaces of enduring and even menacing materiality. At the same time, the city establishes circuits of disposal that become defining routes of renewed consumption, duration, and value. These circuits are invisible, overlooked; yet the remainders that move through these spaces of disposal give Leonia its "definitive form."[1]

Similar daily rituals of consuming and wasting emerge in even greater relief in Cornucopia City, an imagined geography that postwar cultural commentator Vance Packard describes as an example of the furthest extreme of overproduction. In this metropolis, temporary buildings are constructed from papier-mâché, and the factories produce a heap of products that are trucked directly to the dump before they are even able to inundate the consumer market. Through his concocted city, Packard expresses a perceived "crisis of production," a crisis that threatens to saturate markets to such an extent that it overwhelms the possibility for consumption to keep pace.[2] In these cities, disposal, invisible and abundant, is continual and essential to the renewal of production and consumption. Yet there is more to the process and geography of disposal than this loop between production and consumption. As abandoned goods make their outward journeys, they undergo transformations and deformations; they accumulate in peripheral spaces and define well-traveled circuits of disposal. These circuits and spaces of disposal are often hidden, but as Leonia and Cornucopia City imply, they are indispensable to everyday material practices.

This chapter focuses on electronic waste in the form of discarded devices—specifically focusing on the fossilized plastic materials and packaging that house and enable electronics—in order to describe the circuits and spaces of disposal through which abandoned electronics travel. Disposal is not just about garbage trafficked to waste sites, and it involves much more than simply throwing unwanted items in the rubbish bin. Disposal, as it turns out, involves the holding patterns, stockpiling, recycling, and salvaging of materials before they further dissolve or enter another stage of waste. Electronic waste moves not just out of centers of production but also through marginal storage spaces and into recycling depots and, via shipping containers, toward developing countries. In this sense, disposal requires complex infrastructures, practices, and relationships in order to shift devalued objects into spaces for potential revaluation. Such circulations more fully describe the material geographies and practices of disposal, since there is no simple periphery to which objects can be jettisoned. The imagining of the periphery, furthermore, constitutes a topic of investigation: where is there an "outside" to which wastes can travel?

As the previous chapters have indicated, electronics is a rapidly growing industry, with increasing rates of consumption and obsolescence, and for this reason, its waste stream has increased as well. While the exact delineation of what constitutes electronic waste varies, "consumer electronics" of all sorts are scrapped in numbers that are now reaching the billions.[3] Although electronic waste is growing at a rapid rate, the circuits and practices of disposal are not clearly delineated, often because this is a relatively new form of waste. Even with the obvious growth in the number of electronics bought, sold, and discarded, it is actually quite difficult to determine how many of these devices enter the waste stream at any given time, because owners often store and stockpile them for several years beyond their useful life. To further add to the confusion, many countries that export or import electronic waste do not use a specific code to track its delivery, so the trail of disposed devices becomes further obscured in the process of shipping and receiving.[4] The processes and spaces of disposal are not singular but open into expanded geographies. Similar to Leonia and Cornucopia City, the peripheral routes for the disposal and displacement of electronic waste accumulate and congeal into a "definitive," if makeshift, form. This form emerges through disposal practices that are relatively obscured but that are essential in maintaining the apparent immateriality of electronics, even while they are enduring and toxic.

The production of microchips and the screen-based electronic exchanges discussed in the previous chapters, then, extend to wastes generated from electronic production and transmission to consumption and disposal. The focus on consumption here specifically considers how it is continuous with disposal and how consumption patterns can even inform the ability of materials to be "used up." This chapter examines another aspect of digital technology and "use"—not necessarily to concentrate on patterns of interaction between "users" and technology, but to consider instead the more extensive material networks that enable relatively transient forms of "use." But the relationship between consumption and disposal is often neglected. Some studies on consumption suggest that we trace the "social life of things" in order to understand the "trajectories" of commodities.[5] Yet there is a certain difficulty in following "things" in a study on electronic waste. These electronic commodities rapidly expire, have numerous hidden inputs and fallout, and are stockpiled or enter dubious routes of disposal upon their expiration. On many levels, electronic disposal, then, offers a more complete account of electronic consumption, since these technologies have been designed and developed within material cultures of disposability.

Disposal—in the form of use and using up—is a complexly situated process of materialization. To study these material processes specific to electronics, it is useful to account for the multiple "hidden flows" that enable their formation and deformation. Waste is a significant part of the flows of materials that are present not as consumer goods, but as the fallout from production and disposal. Indeed, at any one time, the majority of global material flows are made up of some form of waste. As estimated by the World Resources Institute, these material flows are typically comprised of the by-products and resources that are necessary for the formation of commodities.[6] But these estimates of material flows typically account for the waste generated from production processes and further assume that every item produced will eventually migrate toward consumption and then disposal. Consumption and disposal are protracted spaces and practices that do not necessarily involve a unit-by-unit correspondence. There are vague spaces and processes of expenditure that take place between consumption and disposal.[7] Indeed, a "unit" of consumption does not automatically translate into a unit disposed; rather, consuming, using up, and disposing generate extended spaces of delay, deformation, and demattering. To map these spaces and movements, I take up anthropologist Rudi Colloredo-Mansfeld's suggestion that we should go beyond the social life of things to consider the "social death" of things.[8] This attention to social death can bring to light the extended processes, practices, and places that emerge with the disposal of objects.

In this chapter, I extend this natural history of electronics to encompass the transience and migration of electronics as they pass through and are suspended in circuits of disposal, which cross local and global environments, depend on formal and informal labor economies, and at times require material movements much slower and heavier than the dematerialized networks of electronic markets. To describe these circuits and spaces of disposal, it is also necessary to describe how electronics became so disposable in the first place. As they shift around the globe, disposed electronics sediment as residues from the processes that have contributed to the "throwaway society." This chapter explores how electronics developed within a culture of disposability and how advances in automation, together with new material developments, actually intensified processes of disposability. In particular, the development of plastics played an important role as an ephemeral and disposable material, as well as a material that might be valued for its performance and functionality rather than its durability and solidity. Plastic was, in many respects, the ideal material for the packaging and performance of electronics. As a material composite, plastic further signals the continuity between consumption and disposal, for here is a material that is developed for the purpose of using in order to use up. Plastics and the material technologies of packaging are, then, another critical fossilized fragment and layer to exhume in this natural history of electronics. The ease of disposability, the material transformations of electronics, their consumption and disposal, along with the storing, shipping, and stripping of these technologies—these material practices and spaces together form this account of how electronics turn into waste.

Appliance Theory

During the spring of 2005, in London, a "humanoid" sculpture of electronic proportions loomed seven meters above the river Thames. Composed of refrigerators and computer mice, mobile phones and microwave ovens, computer monitors and washing machines, the three-ton structure represented the amount of electronic waste a typical Briton would generate in his or her lifetime. Five hundred and fifty-three electronic devices in total contributed to the architecture of this sculpture. Yet the number of electronics is as striking as the diversity of devices that now constitute electrical and electronic waste.[9] The pervasiveness of the microchip, as discussed in chapter 1, manifests in an equally pervasive array of electronics and appliances, including everything from irons to vending machines. Many of these devices are more or less "electronic," because microchips and printed circuit boards that channel the flow of electrical currents and information power them. But these microchips are also encased in a skeletal body of plastic and copper, glass and lead.[10] The extended material infrastructures required to house and enable microchips are evident in this motley assortment of plastic appliances. Microchips and plastic assemble into simultaneously pervasive and disposable devices. Leftover electronic devices are primarily composed of plastic and thus appear to be disposable.

The microchip, that miniature conductor and amplifier of electricity, is now neatly sealed in the contours of the everyday. Under the influence of the chip, appliances of all sorts have acquired new "functionalities" and speeds. The ways in which electronics have led to the transformation of objects, materials, and environments may be described as what the now-obscure packaging designer Vernon Fladager has called a "new machine economy." In every such economy," Fladager suggests, and with "every increase in machine speed," new materials, designs, and packages emerge. In fact, "the perfect package material of today can go out the window tomorrow because a new machine economy may make an alternate material a better choice."[11] The electronic package of microchips and plastic is bound up with processes of materialization that can be described through the quickening of matter, proliferation, and increased disposability. This is a machine economy that not only describes altered rates and scales of production but also establishes a temporal mechanism for the disposal of existing materials and designs. Electronics even appear to be programmed for their own elimination, as though an expected part of electronic processing has to do with eventual disposal and erasure.

Electronics, it seems, are prime operators in this transient machine economy. In many respects, electronics are situated within a larger culture of disposability that significantly expanded with the advent of automation after World War II. With automation, there was a general explosion of many consumer goods, which were typically produced to the point of market saturation. New practices of consumption and wasting arose in relation to automated production. In a similar way, practices of electronics consumption and disposal have emerged to facilitate these particular machine economies. Disposability may even constitute an "inventive" use of electronics and peripherals. DVDs have been developed that would expire upon 48 hours after their packages were opened,[12] and certain varieties of mobile phones have been designed to last for only a few days of use.[13] The duration of electronics has dwindled from at least a decade to, in some cases, a matter of hours. Devices appear disposable because they are at once freely available, constantly updated, bound to cycles of fashion, and often increasingly miniature in size. These machine economies encompass more than microchips simply acting on matter. Instead, they evidence the changing material arrangements and practices that sediment within particular technological and material forms. Automation, altered consumption patterns, material developments in the form of plastics, packaging, and shipping technologies and economic geographies have all informed electronic processes of materialization and disposal.

The Throwaway Revolution

The term *throwaway revolution* is used by Packard to describe—if not denounce—the postwar rise in automation and disposability in the United States, when objects with short life spans or limited use increasingly appeared on the market. Technological advancements in automation led to lower production costs, which led, in turn, to a flood of cheap goods on the market, the rise of disposability, and the decline of repair. This was a moment when, as is typically the case now, it became much cheaper to dispose of and replace objects than to repair them. Commenting on the rise of the throwaway revolution within his time, Packard suggests that automation led to an explosion in the number and type of disposable goods available. "Paper plates, cups, bottles, containers have long been disposable," he writes, and "these are now joined, according to a recent report, by 'everything from bikinis to men's blazers, nightwear to student's gowns, curtains to bathmats.'"[14] In Packard's popular critique, economic progress seems to require even more elaborate forms of waste making. If new and improved goods were to be made available and if the economy were to continue to grow, new strategies of consumption and disposal were necessary. Cornucopia City was simply the most ideal—if perverse—installment of this logic: wasting, in the end, stimulates growth.[15]

With automated mass production, a greater store of goods was made available, which enacted changes not just in patterns of consumption but also in patterns of disposal. These changes extended to the availability of a greater variety and volume of disposable goods; yet they also included, as waste theorist Gay Hawkins suggests, "the fundamental logic of the commodity form, seriality."[16] The repetitive production of goods meant they could be easily replaced, old things disposed for new, without any relative concern for where the disposed objects went. Far from constituting a continuation of existing patterns of disposability, the postwar orders of disposability that emerged marked a fundamental shift, not just in the form of commodities, but also in the dynamics whereby they were valued or devalued. Technological advancements that allowed for more rapid product manufacture contributed to the sense that objects were less enduring and more replaceable. Transience and substitution became motivating factors in consumption. This is another aspect of the way in which waste is a generative dynamic, a necessary movement of goods out of consumption-bound circuits and into other circuits of disposal and removal. Practices of consumption become inseparable from practices of disposal.[17]

Disposability is evident not just in the materiality and consumption of goods but also in the growth of the automated production process. As discussed in the previous chapter, with the rise of automation in the mid-twentieth century, changes occurred not just with the gadgets and products available but also to the processes of manufacture and to what it meant to be "automatic."[18] When goods became "electric," they became fluid, moving just as easily from warehouses to markets, homes, and rubbish bins. Matter is programmed—as much for fluidity as for disposability. This stage of automation not only made available a greater abundance of goods but also contributed to the transformation of matter. The accelerated movement of goods was concomitant with a greater sense of dematerialization, plasticity, and disposability. Plastic objects in particular appear to be inscribed with their inevitable movement toward rubbish.[19] These objects tip toward disposal and waste more readily not just because they are more abundant or made of more ephemeral materials but also because they are produced through technologies that enable speed and transience.

The postwar history of technology is a legacy of successively intensifying attempts to electrify objects. Things quicken under the influence of electricity. Once-inert objects transform and are animated by the quiver of electricity. These permutations of matter and electricity corresponded to goods that became more and more transient. The prefix *e-* now potentially can precede even more than markets. The electronic conjoins and augments material and transactions from electronic mail to electronic money and electronic waste. Phones and ovens, cameras and books, leaf blowers and teakettles all submit to the same hazy law of the electronic. Every appliance presents an electrical mutation of an object that once stood still. Matter is charged, but what does it generate? In this general economy of electrification, matter does not just levitate, emanate, conduct, and mobilize; it also circulates, leaks, dematters, disappears, and wastes. The boundaries of objects break down at the same time as they receive an intensifying jolt. Just as the early pioneers from Texas Instruments and Intel anticipated, electronics are now so pervasive that nearly everything is informed in some way by electronic processing. But this pervasiveness is now part of the dilemma, where electronics have proliferated to such a degree that their volume and transience constitutes a material-handling problem.[20] In this "revolution operating on matter" (to quote Serres, cited in chapter 2), electronic objects are produced and designed with increasingly shorter life spans. The effects of increased production and shortening life spans become most evident through the accumulation of electronic waste.

With the rise of automation and electronicization, materials become increasingly indistinguishable from their performance. Materials such as plastic are defined in relation to their functions, as designer Ezio Manzini suggests, from "mechanical function" to "surface quality" to "special electric properties" and even integrating "information input and output systems" into materials.[21] Materials are assessed for their performativity; they are engineered for efficiency, functionality, and, on a certain level, elimination. Objects become smaller, and extraneous components are removed. Function and flow stand in for matter—qualities that are ultimately symptomatic of the electric and the electronic. Matter performs as a package, a surface, a plastic medium for the delivery of function. These operations are also processes of materialization. As discussed in the previous two chapters, electronic technologies enable the capacity for acceleration, proliferation, and destabilization. Yet these same dynamics contribute to the transformation of material and its exchange, as well as the generation of waste and remainder. Electronics and electronicization have as much to do with material developments as with innovations in technology and manufacturing.

Packaging Electronics

It may be the case that electronics owe as much of their development and evolution to the history of plastics as they do to the history of silicon and transistors. It may also be the case that the plastic and the electronic—and, by extension, the plastic and the virtual—have more in common than previously imagined. Plastic is the material that enabled the profusion of disposable packages; it is abundant and pervasive, malleable, and suitable for an infinite variety of uses. But plastics and silicon are also functional materials; they perform operations, so they do more than provide the "raw" material for technologies and objects. These materials in fact inform the possibility of emerging technologies. As "informed material[s],"[22] they exist within processes of materialization and not simply as inert matter.

Informed materials, as discussed by Bernadette Bensaude-Vincent and Isabelle Stengers, operate as more than raw materials, but in fact contribute to the possibility for new technologies and functionalities to emerge. Electronics are comprised of informed materials: silicon enables the flow of electricity and the apparent dematerialization of matter; plastic is inscribed with the capacity for disposability and mass production that now characterizes electronics. Plastic, as a functional material, could be produced in relatively unlimited quantities; it was inexpensive, easily replaced; it could embody the instantly disposable and the imminently possible all at once. As various commentators in the *Modern Packaging Journal* opined, "The biggest thing that's ever happened in molded plastics so far as packaging is concerned is the acceptance of the idea that packages are made to be thrown away."[23] Plastic packaging came to embody all the defining traits of disposability: cheap, abundant, and expendable after a single use. The transience of packaging ultimately contributed to increases in production volumes, where millions of packages eventually grew to billions of packages discarded annually.[24] The single-use purpose of packaging easily extended to all objects made of plastic. Suddenly, not just the casing but entire goods were subject to the logic of abundant, single use.

> Spectacular examples of multi-million unit uses of expendable molded plastics in containers for razor blades, ice cream and other foods, in tomato trays and berry baskets, are demonstrating that a plastic package, while it may be a thing of beauty, need not and should not be a joy forever. Consumers are learning to throw these containers in the trash can as nonchalantly as they would a paper cup—and in that psychology lies the future of molded plastic packaging.[25]

Plastic took the place of paper as the ultimate disposable material, and by doing so, it redefined the material sense of disposability. The rise of plastic packaging was, at one level, part of an effort to minimize the weight of goods previously packaged in glass. The use of plastic in order to minimize associated material, energy, and transport costs was related, then, to a certain drive toward dematerialization. The drive toward dematerialization became continuous with elimination, where goods and packages became more expendable as they required fewer material inputs.

Modernized packaging not only extends to the cellophane and molded polyethylene surrounding tomatoes and soap but also includes the skin around the increasingly transient technological "guts" of machines.[26] In this sense, packaging became a model for disposability that began to inform a whole range of goods, including electrical appliances. Electronics, as with the force of electricity that preceded it, depend on the design of these packages and fluid materialities. Designed packages in the form of electric appliances may enable a sense of efficiency, futurity, and disposability.[27] With electrical appliances and electronics, increasing consumption depended as much on the disposability offered by the package as on the promise the futuristic package presented in the form of technological fashion. Electronics perform in relation to imagined futures; they are packaged in a forward and instantaneous passing of time. Electronics of all sorts have been packaged in ephemeral plastic containers, disposable shells for the conveyance of information.

Plastic, as Roland Barthes writes, "is in essence the stuff of alchemy,"[28] because it enables "the transmutation of matter."[29] So thorough is this transmutation that plastic appears to dematerialize completely in the production process, where it moves from "raw telluric matter" to the "finished object."[30] In this dematerializing movement, which resonates with the electric inventories and immaterial networks discussed in the previous chapter, plastic acquires infinite possibilities for transformation. Any number of objects appear in plastic shells, molds, and packages. Plastic, similar to electronics, mobilizes matter toward apparent invisibility, lending a sense of dematerialization through miniaturization and through accelerating rates of circulation. Plastic is, then, in many ways continuous with the changes enacted by the microchip: these are materials and technologies that emerge as programmed matter, engineered to express malleability, invisibility, and disposability. It was the proliferation of plastics that gave concrete—if immaterial—form to this sense of dematerialization. Plastics are in fact also the material carriers of many seemingly immaterial information and communication media.[31] Just when plastic became so pervasive that it even became the common carrier for electronic technologies, it receded from view. For this reason, plastic partly enabled the sense of virtuality, the sense that digital media somehow operate free from materiality.[32] As discussed earlier, in many ways, immateriality has less to do with the actual removal of matter and more to do with the alteration and "destabilization" of materials.

Indeed, the microchip is a kind of plastic, a reverse packaging that renders malleable the electronics and appliances that it powers. But in fact, these devices take on another level of materiality through electronicization. Objects that were once inert, durable, and relatively benign are now plastic, toxic, disposable, and yet enduring. Electronics do not dematerialize as much as they rematerialize through such (plastic) programming of matter. Plastic, metals, and glass are the primary materials that make up electronics. As the icon of disposability, plastic is part of a group of material composites that often fade from view. These plastic composites constitute what Manzini calls "a world of nameless materials." No longer are objects made of materials that are readily identifiable, such as wood or clay; instead, they are typically composed of a highly engineered and mysterious mix of substances. Computers are assessed less for their material integrity and more for their performance; materially, they may appear at most to be "plasticky" and disposable.[33] What we see with these opaque materials is the operation and image of the devices. Material becomes synonymous with its function and appearance and effaces its own substance. This shift was inevitably aided in large part by plastics. "Plastics have played a fundamental role," Manzini notes, "in triggering the technical, economic, and cultural dynamics that led to the current new scenario of materials."[34] Advances in plastics led not only to the "unrecognizability of materials" but also to the constant redesign of products with materials that promised better performance, with "less matter, less energy, more information."[35] These are the new and nameless materials that dematerialize through the force of information. But when they resurface, they are increasingly difficult to salvage and recycle. Because of the wide variety of plastic composites used in electronics, it is often difficult to sort and recycle these materials for additional use.[36] They are also increasingly troublesome as pollutants and objects that linger indefinitely.

While the electronic industry has speed and turnover in mind, it typically employs materials that last for decades. Here are copper and plastic, mercury and lead, substances thicker and more enduring than any transcription of ones and zeros. Yet for all their endurance, these substances have been essential to the emergence of new orders of ephemerality. Plastic is nearly synonymous with disposability; yet it is also the enduring discardable. Packaging carries with it this deeply ambivalent relation to materiality. Inside the plastic shell that constitutes the predominant material for most electronics are also beryllium, cadmium, and brominated flame retardants.[37] Materials are caught in a tension between the quick and the slow. Ephemerality can only hold at one level; it instead reveals new spaces of permanence. Throw away plastic to discover it lasts for an ice age. The balance of time shifts. The instant plastic package creates new geologies. We now have mountains of congealed carbon polymers. Entirely new landscapes are built up around the fallout from the momentary and the disposable. So this is not just a story about the vaporization of "all that is solid"; rather, it suggests that new forms of solidity—new types of "hardware"—emerge with the program of disposability. Disposability is, then, about more than just overproduction; it also includes conditions of material transience and pliability. Electronic technology may have ephemerality as its guiding agenda, but it unwittingly produces new orders of permanence and new spaces and artifacts of indeterminable duration. The remainders that move through the circuits of disposal, in contrast to the accelerated networks of production and consumption, are drawn into these extended orders of duration and material solidity.

Circuits of Disposal

Disposal and disposability distinctly inform processes of materialization and dematerialization. Disposal and disposability correspond to spaces of removal that stretch beyond singular disposable objects. These are the hidden flows of disposal, involving not just the wasted materials that are used in the manufacture of goods but also the murky spaces where abandoned electronics are dismantled, trafficked, and repurposed. These circuits of disposal reveal how and where these technologies dissolve. The plastic package that encases most electronics has a life beyond its immediate disposal. Indeed, the plastic packaging surrounding electronics enables disposability, a relative sense of immateriality, and mobility. "The distinction between disposability and mobility," as cultural commentator Alvin Toffler notes, "is, from the point of view of the duration of relationships, a thin one."[38] With increasing disposability, goods become so transient that they are rendered liquid and mobile.[39] "Mobile technologies" acquire an expanded meaning, for the most mobile of technologies are, no doubt, often the most disposable. The discards that are mobilized, packaged, and shipped across watery networks give rise to new places and new formations. But where are these circuits and places of disposal?

When we trace through the circuits of disposal, we move closer to what might be Leonia's nebulous boundary between garbage mounds and city. Dirt is supposedly "outside the system."[40] But disposal is about not just attempted elimination but also arranging and ordering, putting aside or situating in relation to networks of exchange.[41] While many studies on waste suggest that garbage is a relationship between "matter in place and matter displaced,"[42] the very process of displacement can, in fact, give rise to places. These places emerge as the residue from attempting to relocate dirt toward an outside. There are many stages and places within disposal, which may extend to sites of storage, reuse, and recycling; transfer stations; and incinerators and landfills. The remainder of this chapter addresses those sites of disposal that are prior to and in transition to the salvage yard and dump, before electronics have reached terminal waste sites (the dump is addressed in a later chapter).

Disposal does not necessarily involve an absolute expelling of unwanted material but, rather, reveals attempts to recuperate or delay the demise of objects in order to postpone their decline of value.[43] Yet the margins where trash is shifted or held are not necessarily sharply delineated but overlap and intersect. Electronics are left on curbsides and in skips, packaged in closets, bundled up in warehouses. These peripheral sites are often actually central but invisible. Part of the process of disposal and displacement involves a willful overlooking of the electronic material debris that surrounds us. Debris lingers in places and often compels us to contend with its dissipated value. A disposed object has, in addition to mobility, a sort of "motility" or stickiness, as geographer Kevin Hetherington notes: objects appear to vanish "only to return again unexpectedly and perhaps in a different place or in a different form."[44] When waste returns and resurfaces, it becomes clear that disposal is about more than matter out of place. Instead, disposal involves a set of practices for dealing with waste (even if this means overlooking it).[45] When we dispose of something, we create places and relations out of the residue of this displacement.

In an attempt to map out these extended spaces of electronic waste disposal, I took a friend's aged personal computer to the nearest recycling facility (at the time, in Montreal). Like many devices of its kind, this PC had sat in a closet gathering dust. Outdated, with a DOS operating system, the petrified machine was a bulky object that one felt should be put to good use but that was no longer functional. As mentioned previously, as much as 75 percent of obsolete electronics are currently stockpiled in the United States.[46] If all the devices that had been stowed away entered the waste stream suddenly, en masse, they would completely overload the system.[47] But there is a good reason why these devices do not unilaterally go in such a direction and why they continue to linger past the point of optimum performance. Not only are the circuits for electronic disposal undefined, but electronics are caught in a set of holding patterns that typifies disposal. The spaces of stockpiling and delay involve sites where "uncertain value" can be assessed.[48] The pause before a more terminal disposal in the dump or before packaging in shipping containers bound for the shores of China and India, is necessary in order to assess the lapsed value of the item. Disposal involves strategies of deferring the moment when objects become rubbish. Electronics initially undergo just such a holding pattern. No doubt, electronics stick around because of the relatively high price paid for them in proportion to the shortness of their useful life. What was at one time a device at the cutting edge of performativity has become an inert black (or beige) box, a device awaiting its final dispatch but remaining in the dim margins.

In my electronics disposal experiment, I located the nearest certified electronics recycler—situated, inevitably, well outside the city center, so that I had to drive the device to its proper waste-handling home. Following this path of disposal, I drove to the near edge of the airport, to a row of nameless light-industrial structures. Numbered loading docks edged up against a continuous plane of corrugated steel architecture, which was interrupted only by the company logo and front entrance. Carting the PC from the car trunk to the front lobby, I noticed that I was the only person in sight, and silent parking lots stretched into the distance. Inside the waiting room, it was clear that this act of singular recycling was unusual, even absurd. I met with the recycler and asked for verification of how the machine would be recycled and if the hard drive could be "wiped" of data (evidence of the success of this process was later sent to me in an e-mail with 13 lines of zeros, indicating no data found).[49] With the recycler's assurances, I handed over the ancient machine, which transferred to the shop floor for disassembly and recycling.[50]

Businesses, institutions, and manufacturers are the primary recyclers of electronics. These groups are often prohibited from sending their electronics to landfills, so they are bound by law to find a recycling option for their machines.[51] While it is not yet illegal in many places for consumers to place their electronics in the trash for eventual shipment to the landfill, more policies now require that electronics are not interred in landfills, as many of the components in these devices are hazardous and present the possibility for environmental damage upon their breakdown and decay.[52] Increasing pressure has also been placed on governments to mandate an "extended producer responsibility," or EPR, that would require electronics manufacturers to take back the devices that they produce, for disposal and treatment.[53] EPR is often seen as a more ideal solution than a mandate that would only require the recycling of electronics, as the latter does not address the fact that the vast majority of electronics collected for recycling are eventually sent, in varying states, to developing countries, where they are processed and handled in relatively unsafe and environmentally unsound conditions.

When we follow electronics beyond their initial disposal, we find that even the apparently final forms of disposal are not nearly so complete and that value is never quite fully exhausted. If we unfold the stages of electronic disposal, we begin to see that there are multiple possible stages of removal, depending on the route that electronics follow. From Montreal to the Bronx and from Pennsylvania to New Jersey, I have visited electronic waste recyclers who have detailed the process of electronics disposal and recycling. Typically, electronics are first collected by recyclers in North America or Europe, who salvage high-grade machines for resale and extract valuable metal from devices for scrap or who alternately bundle defunct machines in shipping containers. In either case, at some stage down the line of processing, the electronics are usually sent to developing countries for scrap and salvaging of components, copper, gold, iron, plastic, nonferrous metals, cables, cathode-ray tubes, printed boards, and more. Raw materials markets thrive on and reincorporate these materials.

The disposal of electronics, then, follows a trajectory between developed and developing countries, where devices migrate from technology-rich regions to those places with an abundance of cheap labor and a high demand for raw materials. While countries such as China are currently regulating against the importation of electronic waste, shipments continue to make their way to Asia, Africa, and other developing countries for recycling and disposal.[54] Using GPS to track the fate of a television recycled in the United Kingdom, Greenpeace activists have mapped how this legitimately recycled electronic device was eventually retrieved in a secondhand market in Nigeria. But there were many stages to locating and recovering the television as it moved across the ocean, from recycler to port, and from port to market.[55] At the same time, many used computers and electronics are sent to developing countries as donations. These devices are meant to contribute to overcoming the "digital divide" by supplying electronics to people who might not otherwise have access to them. Yet the donation of obsolete electronics does not contend with the dilemma that these machines will eventually become waste and will linger in places that often lack the infrastructure for handling these wastes properly.[56]

Indeed, this geographical relation between waste and raw materials is critical to the formation of the "third world."[57] Even when electronics are collected by recyclers in the developed countries, the cost of recycling materials and the geography of markets for raw materials make developing countries a more "viable" place for disposed electronics to be sent in the end. But the cycle of production, consumption, disposal, and recycling is not a machine in perpetual motion, and as the recent collapse in the global market for recyclables suggests, the geographic relationship of manufacturing and waste is not fixed. When developed countries experience slower rates of growth and consumption, the developing countries that supply the products and remove the wastes similarly experience a slackening of activity. During recessions, piles of recyclables stack up in developed countries, as the usual routes for shipping and reusing these materials freeze up. Prices for raw materials can move with the same volatility as apparently abstract indices within electronic markets.[58] Recyclables may even begin to move in new circuits, shifting the relationship between manufacturing and raw materials from more disparate trajectories to nearer geographies; or materials are repurposed not for production but for incineration.[59]

Not only is it often cheaper to send electronic waste across the ocean than to process it locally in places such as North America, but because so much manufacturing takes place in China, the enormous demand for raw materials means the movements of electronic commodity and electronic waste nearly collide with one another, as electronic waste often makes the loop back to the site of its manufacturing.[60] In an account that is reminiscent of Packard's Cornucopia City, journalist Heather Rogers describes how "some shipping companies that bring consumer goods into the United States have taken up rubbish handling. Instead of returning with empty vessels, they fill their cargo containers with U.S. wastes, which they then sell to recycling and disposal operations in their home countries."[61] Shipping containers become part of a veritable conveyor belt, where the movement of goods back and forth across the ocean operates as some well-oiled machinery. Commodity and rubbish anticipate each other. The ease with which these goods move, the lack of distinction between goods for market and goods for disposal, increasingly functions as an abstract system of exchange, as the shipping and receiving of goods now takes place through the automated movement of sea containers.[62] The jumble, reek, and materiality of shipped goods are neatly sealed in containers that do not reveal the contents within. The same containers that ship electronic goods to market could just as likely contain electronic waste: the specificity of these materials has been eclipsed within a standardized container and mechanism of movement.

The majority of electronic waste, then, moves from developed to developing country by ship, which constitutes yet another space of delayed disposal. Electronics that have benefited from advances in plastics, packaging, and automation are then shuttled across the ocean by virtue of this other advance in "packaging." Shipping containers advanced as a maritime technology at the same time that automation and packaging emerged. Shipping containers enabled a new and automated ease of movement, which had a particular influence on the global transfer of cargo. The automated, containerized, and efficient movement of goods by ship resembles those other material, economic, spatial, and temporal changes that were taking place, from plastics to electronics. A technical innovation and newly fluid network of containerized shipping emerged to facilitate the distribution of goods and wastes.[63] These containers move in a liquid and global organization that shifts in relation to cheap labor. Newly discovered peripheries can then become sites for the mobilization and shipment of waste.

Yet within these watery circuits of transport and communication are spaces of material delay. Even at its most routinized, shipping constitutes an extended temporality that undergirds the instantaneous time of electronics. The age of information is more approximate to what artist Alan Sekula calls the "third industrial revolution," which crucially depends not just on electronic technologies but also on these technologies and networks of shipping. While the instant and virtual transport that occurs in digital space often holds sway over our sense of mobility—global, material, or otherwise—in fact, the "forgotten space" of the sea actually enables the movement of most materials, including electronics. So bind-ing are these material flows that they serve as a significant counterpoint to the dematerialized flows of "cyberspace." Sekula writes,

> Large-scale material flows remain intractable. Acceleration is not absolute: the hydrodynamics of large-capacity hulls and the power output of diesel engines set a limit to the speed of cargo ships not far beyond that of the first quarter of this century. It still takes about eight days to cross the Atlantic and about twelve to cross the Pacific. A society of accelerated flows is also in certain key aspects a society of deliberately slow movement.[64]

Electronics and electronic waste trail through these spun-out liquid networks. The suddenness of disposal is drawn out again into orders of material time that are neither plastic nor virtual but, rather, extend into the indeterminable durations of delivery, disassembly, and decay. Just as we position ourselves in the "information revolution," we find that in many ways we are still entrenched in the measured material networks of the Industrial Revolution. In the paused space of shipping, all that had apparently dematerialized rematerializes. Electronics pass through, collect, and sediment in the delay between material registers and in the delay between continents.

Container ships loaded with electronic waste are primarily sent from North America to China, although other circuits of electronic disposal may be traced from Europe to Africa and from Singapore to India. In its report *Exporting Harm,* the Basel Action Network estimates that as much as 50 to 80 percent of electronic waste that is collected in recycling centers in the United States is eventually shipped to locations in developing countries. Guangdong, Lagos, and Delhi receive and distribute used electronics, which move from harbors inland to scrap yards, recycling sites, and resale markets. While electronic waste may have been displaced from one location, it resurfaces in these sites as material for potential reuse and recycling. The question of which "system" is displacing its wastes and how these wastes are configured looms large with the issue of electronic waste. While electronics may have reached the end of their useful life after 18 months in developed countries, becoming "inessential," these same devices are incorporated into other systems where value and use is recuperated and where waste becomes scrap and commodity. These disposed materials are further delayed from complete rubbishing, as they are processed and repurposed in locations often distant from their use and consumption.[65]

What makes electronic waste of particular concern is not just its volume and the fact that it now constitutes the fastest-growing waste stream in developed countries but also that its components are potentially hazardous upon disassembly and decay.[66] The practice of recycling may reinforce a sense that electronics are prepared and processed in a responsible way. But in developing countries, the recycling of electronics occurs through often crude and unsafe methods, including "open burning, acid baths and toxic dumping," which pollute the environment and endanger the workers and local population.[67] Residents in developed countries are relieved of responsibility for these materials, and residents in developing countries process materials and waste that often they did not generate. Perhaps for this reason, sociologist Zygmunt Bauman has spoken of how workers that sort through wastes and recycle materials seem, in the global economy, to also be "disposable people,"[68] expendable and made to deal with wastes from the wealthiest parts of the world. The murky but inevitable relationship between disposability and accountability materializes in concrete form with electronic waste. Circuits for the disposal of electronic waste do not enable its complete elimination; instead, they mobilize these materials toward other sites, forms of labor, and salvage practices.

Dirt, Displacement, Demattering

Recycling may potentially have the effect of increasing or encouraging disposability.[69] Materials may be just as rapidly thrown away, but the sorting, delay, and reintegration of these materials suggests that any problems arising from disposability can be addressed through this reuse. The distinction between recycled matter and rubbish is important in understanding the dynamic of electronic waste and rubbish in general. Recycling is another space of delay within disposal. It draws out materials for sorting, the recuperation of value, and reintegration by transforming rubbish into new commodities. Recycled material can even reenter spaces of exchange and renewed production. In many ways, this transformation takes place through the almost complete devaluation of goods and return to raw material, so that recycled materials move in and out of the economy; they are transformed from commodity to waste and raw material and from raw material into commodity again.[70] But this process involves not just the abstract transformation of materials and values but also the formation of places where material rejection and devaluation takes place. Wire villages, canals flush with broken monitor glass, and alleys full of chemical barrels, which are the typical sites for recycling electronic waste in developing countries, are the actual sites in which these transformations occur. Far from the dematerialized specter of cyberspace, these practices of disposal continually provide evidence of just how material—if dispersed—electronic technologies are.

When we recycle, we repeat the process of delaying the inevitable return to rubbish. Electronic waste may be discarded in one location but then surfaces in another to be processed as goods with marginal scrap value. Yet when that scrap is processed into new electronic components, for instance, it reenters a value system that will mobilize again toward rubbish. Dirt, in other words, is the dynamic.[71] Dirt is, in fact, a constant condition to which objects such as electronics return and against which their value is negotiated.[72] A thing may be reconstituted—as the preceding discussion on plastic reminds—in an infinite number of ways. It may pass into states of disposal and then enter several stages of delay, recuperation, and reentry. When electronics pass through disposal, they undergo such transformations. The displacement of this electronic "dirt" further gives rise to places, social relations, and environmental effects.

It is useful, at the end of this chapter, to return to the earlier discussion on the relevance of approaching consumption through disposal, of understanding the role that consumption plays in using up and dissolving goods and how these practices are guided by the dynamic of dirt. Addressing the interdependent relationship between production and consumption, Marx articulates that "a product becomes a real product only by being consumed" and that "only by decomposing the product does consumption give the product the finishing touch." In this sense, "consumption creates the need for *new* production."[73] Marx's schema creates a loop between production and consumption and focuses on production as the condition to which economies return, where consumption provides the necessary dissolution of products in order to spur new production—hence his phrase "Consumptive production. Productive consumption."[74] While Marx crucially draws attention to the dissolution that characterizes consumption, his analysis does not draw out the spaces and processes of dissolution and does not consider that dissolution may, in fact, be a condition guiding economic exchange. Waste, in this respect, is typically unaccounted for within discussions of production and consumption. Yet waste is a dynamic that influences all phases of economic exchange, providing the basis for the rise and fall of value and the formation of new commodities. Indeed, Marx says as much when he argues,

"Consumption accomplishes the act of production only in completing the product as product by dissolving it."[75] While Marx goes on, in the same passage, to the renewed need for production, a slight interpretive realignment indicates that what is guiding these economic exchanges most of all is the inevitable dissolution of products. Here, products are complete only when wasted. This is a dissolution that occurs not only in consumption but also, by extension, in disposal and the recuperative spaces of recycling.

When we focus on these spaces and processes of dissolution, we can reconsider consumption not only as a process of acquisition but, equally, as a matter of how and where we rid ourselves of objects that are typically manufactured for disposal. Consumption is continuous with using up, and disposal is a critical part of the use of electronics, even if these devices are not in direct control of users. "The issue of de-constitution, of throwing away," archaeologist Gavin Lucas urges, "clearly needs to be related to theories of consumption," because, he suggests, "shedding off possessions can be as complex a process as acquiring them."[76] Consumption emerges not just as a process of dissolution that spurs new production but as a drawn-out process of "dispossession" and "demattering" that critically calls attention to how we get rid of things, how they circulate, where those things go, what residues they leave behind, and what political economies and ecologies they bind together. Disposal provides a way to focus on consumption without eliding this act of using up and without seeing disposal as the simple discarding of matter. Instead, disposal brings into relief those practices, spaces, temporalities, and performances that emerge through the removal and demattering of goods in general and of electronics specifically. Colloredo-Mansfeld argues that "what it actually means to consume an object remains curiously unexamined" and that, in fact, this aspect of consumption as using up is not only a necessary area of study but also reveals how consumption can articulate social relations that "act as generative moments" through expenditure.[77] Consumption and dissolution do not return exclusively to production in this analysis but open up into other spaces that are shaped through the practices and materialities of disposal.

The circuits of disposal discussed in this chapter reveal the locations—often not so officially designated—where the "de-constitution of material culture" takes place.[78] As these practices demonstrate, such demattering is too multilayered and multilocated to occur in any single designated place. If we return to Leonia and Cornucopia City, we arrive this time with a much different sense of the circuits of disposal in these places. Cornucopia City trucks its goods from the production line to the dump: it does not account for the necessary role of consumption in using up goods and extending disposal into multiple places. Leonia simply shifts its continually discarded goods outward, to an unnamed margin, which could just as well be some electronic waste dump in Guangdong. We not only need places of demattering; we already have them. They just tend not to register as places of regard. But these places of disposal continue to exceed their boundaries, forcing us to reconcile ourselves to the effects of our wastes—electronic and otherwise. Yet there are also spaces of more official demattering that we can turn to in order to consider how we deal with the loss of material culture. The museum or archive is perhaps primary among these designated spaces for witnessing or arresting the erosion and erasure of material culture. These are sites that manage the duration and space of material release but also preserve a concrete record of the program of transience within electronics. In the next chapter, I consider how the museum and archive offer up spaces of demattering and disposal, as well as material memories of failed technologies.

Shipping containers in Singapore port, 2006. (Photograph by author.)

Electronics at a Montréal reuse and recycling center, 2004. (Photograph by author.)

FOUR

Museum of Failure

THE MUTABILITY OF ELECTRONIC MEMORY

Computers offer an interesting daydream: that we may be able to store things digitally instead of physically. In other words, turn the libraries to digital storage; digitize paintings and photographs; even digitize the genetic codes of animals, so that species can be restored at future dates.

—TED NELSON, *Computer Lib/Dream Machines*

The possibility will arise that technics, far from being merely in time, properly constitutes time.

—BERNARD STIEGLER, *Technics and Time*

Refuse of History

In the Computer History Museum in Mountain View, California, a veritable warehouse of machinery is on display. Here are a Jacquard loom and Hollerith punched card machine, the Cray 7600 supercomputer and the JOHNNIAC. Many of the machines are notable for the contributions they made to the development of computing; others are representative examples of everyday electronics from a particular era. Yet all of these machines, regardless of merit or extent of distribution, are silent. Disconnected and unplugged, the devices seem to meditate under a layer of dust, which is amplified by the fluorescent lights. In this hall of exhibit placards and mute machines, other features slowly begin to rise to the surface. One mainframe, the curator tells a group of visitors, has peculiar markings to which he would like to draw our attention. The machine is the WISC, or Wisconsin Integrally Synchronized Computer, which was developed between 1951 and 1955 as part of the PhD thesis of Gene Amdahl at the University of Wisconsin. While this machine was pioneer-

WISC mainframe, Computer History Museum, Mountain View, California, 2005. (Photograph by author.)

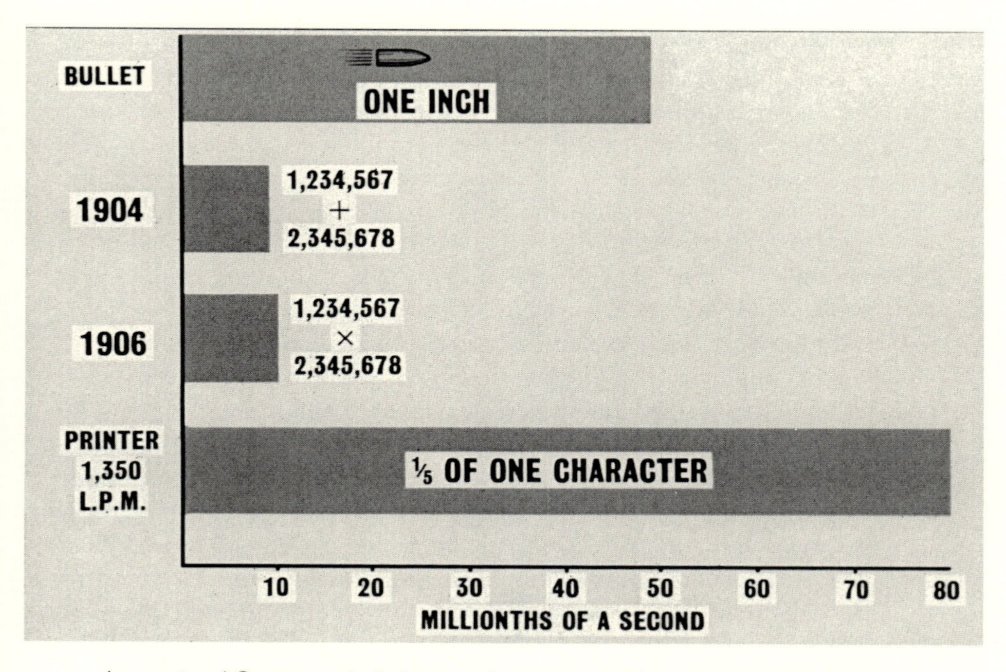

International Computers Ltd. diagram of computing and printing speeds, ca. 1970, Science Museum of London. (Courtesy of Fujitsu.)

ing for its time, it quickly became obsolete as many new mainframes entered the market, and Amdahl went on to develop other computers at IBM. The markings that we are directed to examine are a scattering of bullet holes across the console of the machine. According to computing legend, when Amdahl moved on to IBM, the device was used as a training computer, only to be later retired and moved to a professor's midwestern basement. In this subterranean storage space, it became the direct or indirect object of rifle target practice. Once it was eventually rescued and preserved in the Computer History Museum, it bore these indelible marks of its other life, when it once lingered in a state of disuse. The holes that puncture through the WISC's bullet-riddled console aim into the secret workings of the machine. Memory drum and electric circuits are not the only apparatuses that lie behind its opaque exterior, however. These bullet holes also tear into the mechanics of technological obsolescence. They are a reminder that in the endless tale of technical evolution, electronic machines are regularly cast aside, become obsolete, and are kept in storage as inert remainders. Before it entered the museum, the WISC acquired this other layer of dust, a rough grain recording the fate of failed electronics.

In the museum and archive,[1] there are failed and obsolete technologies in abundance. On display are objects that at one time were so terrifyingly new they seemed to tip into impossible future imaginings. But the objects lapse into disrepair; they fail to remain new forever. There is always a perceived need for another upgrade and another, ad infinitum. Cultural theorist Will Straw suggests that "the sites in which unwanted cultural commodities (old records, books, etc.) accumulate are, at one level, museums of failure."[2] Any museum or archive in which electronics are held is a collection of repeated obsolescence and breakdown. But failure is only one part of this story. Whether in a state of decay or preservation, obsolete devices begin to express tales that are about something other than technical evolution. By tearing into the mechanics of obsolescence, the WISC bullet holes do more than simply reveal the failure and mutability of machines. Above and beyond this, the bullet holes open into another order of time that exceeds the trajectory of progress and innovation. Obsolete commodities and technologies, as Benjamin explains, open up other orders of time by falling out of the time of progress.[3] Instead of demonstrating historical advances, these objects provide evidence of the dust that sediments as a record of these material and technological imaginings.

There is yet another image of bullets in this history of electronics. A chart by International Computers Ltd., or ICL, a now-defunct company from the United Kingdom, compares the speed of bullets to the speeds at which mainframes process or printers output data. Within the usual pronouncements on the progress, speed, abundance, and overload of new technologies, the dust is most often overlooked. But dust may, in fact, be a more accurate gauge of these technological objects. For all the successive doubling of computing speed and for all the flurry of new electronic innovations, there is a corresponding degree of electronic obsolescence. While electronics may seem to demarcate the accelerating speed of information, they also uncover the accumulation of dust. The speed and effect of "progress" has a necessary remainder. But if we suspend the assumption of progress and concentrate on the discarded objects, we can begin to consider how dust may be an underlying condition. Nowhere does this become more apparent than in the museum and archive. The attempt to preserve electronics collides with the fact that these are machines programmed for their own destruction. Such a collision reveals economies of electronic time that not only are problematic for the archive but also undo the narratives of speed and progress so central to electronic technologies. This chapter considers how archives shift under the influence of electronic temporalities. Electronic memories—as electronic fossils that both settle into the form of hard drives and storage devices, and that scatter through operating systems and archives alike—give rise to specific modes of electronic waste. It is these fossils of electronic memory that this chapter investigates.

The WISC stands among devices that were once novel inventions but are now arcane and relatively impenetrable artifacts. The uses and legibility of these devices have passed. They are forgotten technologies. But in this space of lapsed function and memory, the devices persist as remnants. As suggestive remainders, they become newly resonant. Plastic cases of robot eyes vacantly fix on some far distance, tangled wires mass together as though these metallic devices were born of aquatic origins, and video game consoles resemble hungry industrial ovens. Impenetrable or strange, inaccessible and decaying, beyond the reach of function and so made bizarre, but ultimately engendering new imaginings, these objects undergo an electronic alchemy that gives rise to the unexpected. In fact, the fantastic aspects of technologies, as Benjamin suggests, are revealed in both their making and their breaking. These are the two moments when the utopic future that technological objects promise is revealed. When technologies become obsolete, we have the opportunity to reexamine these utopian promises and to recast the material, political, and historical terms on which we encounter these devices.[4] Were these electronics, strange forms to us now, once meant to transport us to some utopian condition? Their earnestness suggests as much. They are nothing less than molded plastic epics. Yet the rush of innovation congeals into a fossil record. The short life and quick death of these objects settles into a layer of this natural history that reveals the critical relation between temporality and materiality and between progress and obsolescence.

While the attention of many writers on culture and technology returns to the successive "creation," or the next "paradigm shift," Benjamin suggests, instead, that we attend to these orphaned objects and places. From them, he generates this particular form of natural history, in which history sediments into things. In this method, knowledge settles into dead objects through mortification.[5] The *mors,* the degradation that comes with decay, of falling out of favor, is not a common topic in the world of electronics. But still, we have the leftover shells, and these are not without their substance. It may be that from this state of obsolescence, it is possible to learn the most about what these technologies promised and what fate befell them. Failure presents the fossils of forgotten dreams, the residue of collapsed utopias, and the program of obsolescence. Through the outmoded, it is possible to move beyond those more "totalizing" aspects of technology, such as progress, teleological reasoning, or the heroism of invention.

Outmoded technologies reveal the unintended and residual, and they allow access to these other registers, spaces from which also issue the mythic, the failed imaginings, and the alchemy of electronic devices. The dustbin of history, the refuse of history, adds up to a much more dynamic record. Waste renders problematic the telling of history as an unwavering narrative of progress. Past objects do not illuminate the past as much as reveal the inevitability of decay, or "irresistible decay," as Benjamin terms it.[6] But disintegration and decomposition are not the dystopic angle on utopic promises; instead, they offer up a way of characterizing processes of materialization that are not simply causal or informed by ideological objectives. Histories and material cultures are not immune to decay, and they may even engender a more intensified relation to it. It is to such a transformation that this chapter turns in order to investigate an electronic alchemy that is not about automatic progress as much as the complex ways in which machines fall apart.

Some of the best places to witness the unwitting decay of electronics are in the very spaces where they would be preserved. Many electronics relegated to museums undergo such a rapid scale and rate of dematering that preservation is rendered problematic. *Preservation* becomes another word for managed decay, for a delay within the extended process of disposal. The museum may also be construed as a space of disposal.[7] Often, the museum and archive collect and stow away objects that have for most purposes been disposed of and removed from the spaces of everyday circulation. The museum collects objects in storage, much the same as the electronics lingering in closets, attics, and warehouses; but the objects in the museum must be continually sorted and deaccessioned in order to make way for new objects. Moreover, the migration of archived materials to digital formats has shortened the life of most museum objects, tied as they are to the life of electronic data. These newly digitized objects require, in turn, a continual transference, updating, and migration to newer formats in order not to dissolve into the inaccessible static of obsolete electronic data. Electronic archives, electronic memory, and electronic waste are bound up in these shared processes of materialization; they are part of the same constellation of data and dream.

Electronic Memories

At first glance, electronic memory and storage seem to be the ideal instruments for an enhanced process of archivization.[8] With electronically assisted memories, it is possible to process, store, and transmit more material and data. Archives promise to be nothing less than advanced and virtual versions of the Library of Alexandria. The introduction of the integrated circuit is often considered a significant marker in this revolution in the processing of electronic memory, where chips introduced in the 1970s initially had one kilobit of binary storage and have since grown to several gigabytes of temporary storage (and counting).[9] Electronic memory repeatedly realigns toward expanded volumes and velocities.[10] Memory storage and memory processing (or RAM) have both increased appreciably. The sheer increase in the scope of memory processing and storage generates, in turn, the need for new terms and concepts to describe these altered temporalities and materialities. Such memory capacities have even been described as "global," processes that occur through extended networks and so are beyond the scope of any "earth-bound body."[11] The scope of such memory has gone beyond being a mere material extension or surrogate, to become a seemingly independent entity. Computers synthesize "times"; they generate temporalities and memories. But these temporalities are distributed and extend beyond any single human capacity.

The operation of memory reaches such an extent that it has, at various times, even seemed to render "man" obsolete. In 1962, for example, as Arthur Clarke writes in "The Obsolescence of Man," "Marquardt Corporation's Astro Division had just announced a new memory storage device that could store inside a six-foot cube *all information recorded during the last 10,000 years.*"[12] Clarke describes an unprecedented archive, which occupies a mere six cubic feet in physical space but extends out 10,000 years into the temporal dimension. He elaborates on this new temporal compression,

> That would mean, of course, not only every book ever printed, but *everything* ever written in *any* language on paper, papyrus, parchment or stone. It represents a capacity untold millions of times greater than that of a single human memory, and though there is a mighty gulf between merely storing information and thinking creatively—the Library of Congress has never written a book—it does indicate that mechanical brains of enormous power could be very small in physical size.[13]

With this perfect archive, "man" (as embodied memory) effectively becomes obsolete. Such a lumbering, bulky, and inefficiently material memory is no match for those cubes and circuits that can cut through entire generations with the flick of a switch. These memories can also no longer be described as surrogate.[14] Instead, they are planetary, even cosmic. The computer not only synthesizes times; it consumes them. In this respect, it may be useful to get a gauge on what the machine's "diet" entails. Here is a machine capable of devouring centuries, of processing comparisons across epochs. Its extension across time makes electronic technology appear to be the ideal mechanism for a more advanced archival project. But this extended memory is not without its alchemy.

In the first chapter of this book, we encountered Bush's proposal for a "Memex," a device that would help people organize and retrieve mountains of data (and so stave off potential overload). The Memex, as previously discussed, would aid in this process by compressing a large store of information to a minute size, where reams of papers and entire sets of encyclopedias could be accessed within the space of a tidy desk. The principal use of the Memex, of course, is as an aid to memory. It is an archival device and allows for ease of storage and retrieval, cross-referencing and association. It ultimately improves our ability to "get at the record." With untold storage space and an appetite to consume anything, the Memex would dine on books and records, letters and photographs. Its digestion of this material, stored away for ready access, would be aided by its "mechanization," which would allow material to "be consulted to exceeding speed and flexibility." The Memex is, then, "an enlarged intimate supplement" to memory. This is a memory machine that depends for its usefulness on consisting in large part of "mechanism," or, in other words, of the means to process, access, and deliver information that might otherwise have disappeared from memory.[15] But what is presented at first glance as an aid to memory quickly becomes an entirely new order of memory, time, and processing.

Bush's proposal dates from 1945. But in 2001, Microsoft took up the Memex proposal as an opportunity to develop a new, similar program: MyLifeBits. This modern-day attempt to implement the Memex proposes to "encode, store, and allow easy access to all of a person's information for personal and professional use." When this project proposes to work with "all of a person's information," it literally intends to catalog *everything*, including "articles, books, music, photos, and video," together with all that is "born digital," including "office documents, email, [and] digital photos."[16] This is an archive from which nothing escapes. Furthermore, new material continually presents itself as worthy of recording. Cameras with sensors may even document "environmental information" by taking continual snapshots, archiving up to 1,000 images per day. The drive to archive everything even begins to burnish entire centuries with a particular grain, where the twentieth century will have a much different resolution than the twenty-first century. Indeed, the authors speculate that "21st century users may expect to record their life more extensively and in higher fidelity—and may drive a market for much greater storage."[17] These increased resolutions and quantities also mean that the scope of "multimedia" exceeds all imagination. This "transaction processing system" is capable of capturing "virtually everything in a person's life at meaningful resolution—user's interaction with others, as well as logging location, calories, heart rate, temperature, steps taken, web pages, mouse clicks, and heart beats."[18] The volume and type of material to be indexed is quite simply "inexhaustible." Even the expiration date on the milk in the refrigerator can be made into archivable, searchable, and programmable content.

While one leap occurs within this project when all media are digitized, yet another occurs when the entire world is rendered as potential digital media-in-waiting. Why should our heartbeats not be stored and accessed as digital traces? Not only does the digital operate as a device for managing other media; it also permits the ability to operate on that media.[19] Archivable data is calculable data. This is the other critical component of electronic memory: not only does it store, but it also programs material for operation. The computer is the universal machine.[20] It can operate on anything as long as that material is rendered in digital format. The consequences of digitalization are seldom mentioned. While these electronic mechanisms may seem to preserve "endangered things" in a relatively permanent archive, they also present the dilemma that, as Kittler notes, "the medium that archives all media cannot archive itself."[21] As we input heaps of data into digital devices, it seldom occurs to us that the digital devices themselves are rapidly changing entities and that they, too, generate data for the record. Moreover, the inability to archive itself means the electronic mechanism has a fundamental inattention to its own temporal configuration.

When memory is apparently separated from material requirements, compressed as it is within a compact processor, the course of time not only computes in much different ways but also variously comes to ruin. The electronic archive grows to prodigious proportions, yet this same archive may be completely inaccessible in less than a decade unless it is reformatted to keep pace with new electronic technologies. Ten thousand years may be ensconced in a six-foot cube, but without a means to access the data, we can only gaze wistfully at the minimal cube and wonder at the inaccessible 10,000 years that did, at one time, fire through its busy circuits. Increasingly, this issue has become a quandary for electronic archives. Former director of the Getty Conservation Institute Miguel Angel Corzo indicates how digital media of even the most significant cultural moments quickly evaporate. "For instance," he writes, "digitized images from the historic 1976 Viking mission to Mars that had been carefully stored and appeared to be in good condition are now degraded and unreadable."[22] As the MyLifeBits researchers note, the primary dilemma that their proposal encounters is the problem of longevity; in other words, "how do you insure that your bits will live forever and be interpretable?"[23] The upgrading of hardware, the introduction of new operating systems, the transience of data formats—these elements are constants within the development of electronics.[24] Each wave of new and improved tools of electronic memory potentially will obliterate past records and render them inaccessible, unless, of course, we rerecord everything in this new format.[25] The only perceived feasible solution, then, is to develop "emulation systems"[26] that will move data to new platforms at least every ten years, in order to ensure that we can still "get at the record." Yet even this hopeful process of erasure and reinscription typically falls outside the archival record. Is it possible that these electronic archives are at once the most extended temporal registers while simultaneously having the shortest duration of all archives to date?

Much more than digital media comes into play when we consider all of the possible elements of electronic systems that may, at any moment, become inaccessible, incompatible, or obsolete. The failure of components or seemingly isolated objects may actually reveal the systems to which objects are connected, because these elements are, as sociologist Harvey Molotch notes, "'interactively stabilized' practices and things."[27] Any archive that attempts to preserve electronic objects enrolls itself inadvertly in the preservation of electronic systems. But the project of preserving electronic systems makes previous forms of preservation—from objects to paper, with their threatening worms and mildew—pale in comparison. "Basically," as Sterling writes, "you the lonely archivist are trying to support and preserve an entire cybernetic post-industrial system."[28] This system presents an untold number of pitfalls not only due to the quantity of objects to be preserved but also due to the infinite possibilities for failure in the preservation project. The scope of these breakdowns means that "the central processing chip can fail. The operating system can fail. The language that supports the operating system may be discontinued and no longer supported."[29] But these failures only begin to hint at the full scope of possible disasters. Indeed, as Sterling writes, "it gets worse."

> You may lose the subtler forms of adjunct software, such as the screen display software, the printer drivers, the audio chips. The keyboard format may not work. The application may fail. The data storage formats for the application may no longer be supported. You may have different screen dimensions, or different graphics formats that fail to display for various bulky, difficult, inexplicable reasons. The material you are trying to preserve may

be encrypted. The key may have been lost. There may be digital rights management difficulties that forbid copying. And, the storage media themselves are physically unstable.[30]

At the same time that we are stuffing nearly everything into electronic formats, those same formats prove to be incredibly short-lived and contingent. When we are presented with the possibility of devising a database "for life," we must confront the reality that there is no digital format that has proven to be capable of such an inordinately long time span. Even now, five-inch "floppy" discs that are no more than 10 years old are typically unreadable, as there are very few remaining machines capable of extracting data in this format. From software to operating systems, Web sites and storage media, electronic technologies collide in multiple layers of transience. The universal archiving medium, electronics, cannot itself be archived. So much for the database for life. Rather than gesture toward the permanent and enduring, however, perhaps we should address more fully the thoroughgoing transience of the electronic.

Electronic time is fleeting. With this realization, we can dispense with narratives of enduring cultural records and instead begin to study this technology in its volatility. Here is a technology that would archive everything and even transform nonmedia into digital format. It can store and sort and search and process beyond measure, and it can erase all of this data with a silent and swift system collapse. The promise of absolute memory, of a record of everything, gives way to erasure. But in this dynamic, we can begin to uncover the temporal economy of these electronic memory technologies. Electronic archives depend as much on erasure and transmission, it turns out, as on storage.[31] Because the archive is more akin to a network than a storage shed, the archive is most effective when its contents translate into transmission, into the ready execution of programs. Memory, in this respect, always occurs as a kind of program. Storage does very little on its own. How many inaccessible hard drives from decades past can one stare at before realizing such a fact? Without a means of "getting at the record," these electronic devices are little more than doorstops. But with the "program" of memory, no item in storage is left idle for long—in fact, the longer it is left idle, the more chance there is that it will disappear from the record completely. Instead, the distance between memory and real time nearly collapses.[32] In such a temporal economy, memory operates in a much more immediate and instantaneous way. Storage is only accessible to memory if it, too, moves through these rapid transmissions.

We can see how assumptions regarding the archive—whether electronic or otherwise—founder when they fail to consider how memory, duration, and times emerge through technologies.[33] Electronics may have given us the nanosecond, but they have also given us digital decay. The latter is much less about placing ourselves on a known—even if imperceptible—timescale and much more about a set of unfolding temporal effects. Indeed, digital decay can be so disorienting that it may be difficult to gauge whether the trash is coming or going, whether the rubbish is in the past or surfaces as a sure marker of the future. Sterling suggests that in these technologies of time, we repeatedly generate leftovers, all sorts of "prehistoric" hardware. In this sense, "trash is always our premier cultural export to the future."[34] Ancient hardware turns up in the future, but, then, what would the future be without its rubbish? Surely it would lose all sense of futurity—of newness—if it did not have some identifiably obsolete remnants. It may be that electronic technologies do not just generate obsolete remainders but also positively rely on these remainders—these old media—to gauge what is new. It is in this same temporal density that emerges with obsolescence that Benjamin is able to direct us to let the dust settle until we see that the driving force of technological progress may, in fact, be standing still.

Programmed for Obsolescence

As is apparent by now, the history of postwar computing is full of tales of obsolescence. Countless electronic artifacts could be selected as evidence of the ways in which innovation turns to ruin. Even projects that would attempt to document and analyze current and historic developments in computing fail. The U.S. National Bureau of Standards attempted to catalog all the extant computers in 1951 with its report "Evaluation of Automatic Computing Machines." The bureau failed in its survey and had to abandon the project because too many machines were developed far too quickly to document.[35] Similarly, in the preface to the second edition of his study *A History of Modern Computing*, computing historian Paul Ceruzzi indicates that just as he was completing his manuscript, he felt it was rendered obsolete by ongoing developments in computing.[36] He suggests that new orders of time continually emerge that may explain this dilemma of never being able to capture the world of computing and electronics.[37]

The historian will never manage to compile a complete account of computing, because this is an ever-shifting and rapidly accelerating field. The speed with which innovations occur means that the speed of analysis and capture is seemingly too slow to keep pace. If everything we write about electronics becomes obsolete the moment we put pen to paper (itself an obsolete turn of phrase), then perhaps we should begin to address this dynamic of obsolescence. Clearly, this is the one thing that does not fall out of fashion. We can count on the dynamic of obsolescence to retain relevance not only with Babbage's Difference Engine of 1822 but also with the seemingly futuristic "Internet of Things." Histories and machines alike are obsolete the moment they are introduced. These are not tales that will be fixed or definitive. But, then, does this not require that we begin to reconsider histories of electronics as histories of expiration? There is certain impossibility to writing the "now" of these technologies, a fact that is only made more evident by the strange persistence of the self-defeating term *new media*.[38] The new, with such rapid rates of innovation, is inevitably always old in a very short amount of time. Can we even refer to currently new media as new anymore? The Internet seems positively prehistoric, having been in common use for more than a decade now. This may explain why researchers have declared the death of the Internet, or the advancement of the Internet, or the rise of Web 2.0. In attempting to capture a technology so driven to outdo itself, the very "histories" that would describe it must turn to the dynamic of obsolescence and to an understanding of how transience forces a reevaluation of those histories.

Of all the types of obsolescence, technological obsolescence often appears to be the most incontrovertible. Technological advances present an inescapable logic for upgrading and discarding. Packard, who attacked obsolescence as a strategy parallel to disposability, suggested that it was adopted by marketing experts to deal with the problem of overproduction and underconsumption. From his perspective, commodities—particularly technological commodities—began to be produced with rapidly diminishing expiration dates. They were subject to "planned failure," which ensured that consumers would always have a reason—whether through the desire for the new or through mechanical breakdown—to ingest more commodities.[39] Unlike Packard, Toffler attributed such rapid advances not to intentional manipulation on the part of producers but, rather, to advances in technology, of which the computer was a prime example. The new machines were simply better than the older versions.[40] This dynamic was not something that could be "attributed to the evil design of a few contemporary hucksters" but, instead, stemmed from the rapid rate of obsolescence, the "fantastic rate of turnover of the products in our lives," which was yet another sign of the "entire accelerative process—a process involving not merely the life span of sparkplugs, but of whole societies."[41] For Toffler, obsolescence did not exclusively proceed through advertisers or industrial designers pulling some imaginary puppet strings.[42] Instead, he suggested we were all subject to these forces, which had become larger than any single person or organization.

This debate and the possibility that obsolescence is integral to modern production direct us to the larger scope of technological obsolescence, which encompasses not just the regular introduction of new gadgets but also a seemingly involuntary impulse. Technological obsolescence seems to proceed automatically, without need for overarching control, because machines are programmed for failure. These machines are self-propagating and self-obsolescing. Their obsolescence is literally built-in.[43] For these same reasons, technological obsolescence is continuous, as many writers suggest, with the program of human obsolescence.[44] Technological obsolescence finally makes humans obsolete as the directors, the replicators, and the saboteurs of machines.[45]

Obsolescence appears to be "built-in" on multiple levels, from the actual decay of hardware, software, and content; to the economic requirement for continued innovation; to the way in which the pastness and the newness of electronic media and technology is narrated. Technology even acts as a reference point for change, where not to be at the technological forefront is a sure indicator of obsolescence. "Obsolescence," Sterling writes, "is innovation in reverse."[46] But this pairing may be more coincidental than causal. Obsolescence plays a role in validating innovation; without obsolete objects and technologies, we would have no register of what constitutes an "innovation." Obsolescence is not just that which is left behind but also that which persists in the present as a discernible marker of disuse. In this "production of obsolescence," as cultural theorist Evan Watkins writes, "'yesterday's' innovation" does not simply disappear; instead, it endures so that other, new technologies may appear to be innovative all over again.[47] New technologies seem to be innovative, particularly in contrast to those rusty cogs and sprockets that surround us or to those defunct electronics that are shipped for recycling to developing countries. Obsolete technologies do not disappear into the past so much as they shore up the margins, playing silent witness to the newness of the newest devices. Obsolete objects continue to play a role in the overall market of change; they reveal the "capitalization of change."[48] Obsolete objects, the "day-before-yesterday's technology," are integral both in producing the future and in producing the "primitive," the uneconomical, the passé.[49] Museums and archives that collect obsolete electronics play a fundamental role in validating the newness of the latest innovations. Obsolescence is not so much innovation in reverse as it is the ongoing maintenance of a sense of technological development. Without this rubbish, which is coextensive with new technologies, we would not have a sure indicator of the progress we have made. All around us are the machines that readily propel us into the future. To move us ever forward, many of these machines do not even need to function.

Fad Machines

Most electronics have a longer presence as defunct remnants than as fully functioning, plugged-in and systems-based technologies. But such a temporal inversion is almost to be expected when the rate of innovation within electronics has contracted to as little as 18 months. Moore's Law has been the benchmark by which computing revolutions have been measured since 1965.[50] A near golden law within the world of computing, this may even be "the true driving force of history," as argued by Ceruzzi as well as engineers in the industry.[51] The computing revolution, then, depends on the appearance of new technologies every two years or less in order to materialize as devices that seem to be "revolutionary."[52] The revolutionary potential of electronic technologies can be measured literally—by the frequent revolutions, the successive turnover in devices.[53] Yet the driving force that writes orders of computing history is also a force of transience. With every new set of technologies, the devices currently in use edge even closer to obsolescence and become even more likely to fail, whether through incompatibility or lack of repair. These "revolutions" ebb and flow with regular predictability. Yet just how do these revolutions become so consistently executed? To what extent is Moore's Law a driving force, and to what extent has it come to be a "self-fulfilling prophecy" through vast, if transient, infrastructures and investments?

Sociologist Donald MacKenzie uses the phrase "self-fulfilling prophecy" to refer to the ways in which technological growth or failure is shored up by expectations and investments that ensure such a performance. "Moore's Law," he writes, "is not merely an after-the-fact empirical description of processes of change in microelectronics; it is a belief that has become self-fulfilling by guiding the technological and investment choices of those involved."[54] While the belief in technological growth may "be dashed as technologies encounter the obduracy of both the physical and the social world,"[55] a lot of effort goes into attempting to make these guiding principles come to fruition. To maintain facilities advanced enough to fulfill the objectives of Moore's Law, Intel regularly updates its chip fabrication facilities, amortizing as much as one billion U.S. dollars per year just in production costs. At this rate, Gordon Moore himself has noted that this means Intel factory facilities are completely replenished every four to five years.[56] To remain at the forefront of technological change, investment must be made in infrastructures that enable these rates of change. As MacKenzie aptly states, "Persistent patterns of technological change are persistent in part because technologists and others believe they will be persistent."[57] Increases in computing become a guiding factor as much as an expected development for the computing and electronics industry. This could be described as a more than empirical phenomenon: the evidence of computing growth is bound up with technological imaginaries and ideal rates of advance. Processes of materialization, which span from fabs to expressions of innovation, maintain and stabilize these more-than-empirical events.

To maintain the rate of innovation set by Moore's Law, any number of devices are deployed in order to arrive at advanced computing speeds, including renewed factory facilities as well as modified components, new chemical combinations, and experimental technologies. While the requirement exists to maintain the standard of Moore's Law, this rate of technological growth also serves as an ideal guide for maximum growth. It is actually riskier to attempt "optimization" of speeds, as it presents "the risk of technological failure."[58] We have the self-fulfilling prophecy of Moore's Law, the basis for a new "revolution" in computing and electronics every 18 months. This, as much as the number of transistors on a microchip, is the basis for "a technological trajectory," or an "institution" for technological developments.[59] While it appears to be automatic, a force emanating from technology itself, Moore's Law is, in many ways, bound up with sustained efforts to maintain the regularity of this change. The landscape of electronic transience consists of more than just the sudden and magical, if regularly predictable, technological innovation of its own accord. This level of innovation is an industry standard, and it is the rate of change to which any number of social, political, economic, and technological dials are tuned. Ideal levels of innovation may fail to materialize, but these trajectories describe the processes whereby new machine economies emerge together with new machines. A natural history of electronics, then, encompasses not just the marvels of electrical firing and decay but also the extended systems and resources established to underpin the innovation and turnover of electronics.

When Toffler refers to the "fad machine," to discuss the regular turnover of objects, he suggests that transience has developed to such a point that we have entered what he calls "the economics of impermanence," where products are built for the short term.[60] Collapsing duration, ephemerality and transience are, as he argues, distinctly enabled by advances in technology.[61] It perhaps comes as no surprise that Toffler would cite "automation expert John Diebold" as a voice commenting on the need to think of products for the short term. So insistent is the fad machine that it seems to be automatic, to constitute an automation program. In the late 1960s, thousands of products were developed and faded away in rapid succession. When products at one time may have been in the market for several decades, increasingly they were present only for a matter of months and, at times, weeks.[62] A side effect of such production and obsolescence is the colossal amount of rubbish that accumulates, as products are discarded and substituted for newer models. Yet all of this debris is fodder for the archive. With such a prolific program for the production and consumption of goods, how does the archive—as both a culture and a technology—shift to contain and sort the remains of everyday life?

Salvaging Archives

From the "economics of impermanence," we arrive at the archive of impermanence. Today, the archive must contend with the dilemma of preserving self-erasing artifacts, of fixing a material culture that is intensely ephemeral. The move to archive "everyday life" has led to the reinvention of the project of preservation, where almost everything constitutes possibly archivable material.[63] But in the attempt to archive everything, we would encounter the everyday and its "distorted memories" all over again. With such all-encompassing means of archiving at our disposal, we are able to store everything, but in that ambitious documentation, we at the same time inevitably include the decay and oblivion that, at one time, it was the task of the archive to guard against.

Electronic storage brings the tension between memory and oblivion to a renewed collision. Electronics may have even contributed to rewriting the archive's program. The transience and even banality that emerge with electronic storage extends to new levels, where heartbeats and expiring milk acquire a place as archive-worthy data. In fact, through the monumental task of archiving everything, the archive becomes more akin to a disorderly waste site, which then requires processes of computation to make sense of the welter of material and data. This seeming contradiction is the functional basis for the electronic archive, where material is digitized in such quantities so as to appear chaotic, yet the engines of computation can, at the same time, search and process this material toward order. This capacity for at once creating and transforming waste may even change the "memory of waste" referred to by media theorist Wolfgang Ernst.[64]

> What will happen to the "memory of waste" under conditions of electronic storage, when the hermeneutic instrument for differentiating between value and rubbish in things to be stored is abolished in order to make place for a cybernetic register of non-hierarchical hypertexts. The electronic age succeeds in erasing the opposition between monumental inscription and discursive flow.[65]

When the memory of waste shifts, so, too, does memory itself. The criteria for distinguishing significant event from everyday detail collapse with the ability to store everything without distinction and to process the material according to real-time requirements. In this respect, Ernst writes, "*memory* is being transformed cybernetically into synchronic information networks."[66] The relevance of particular material as archive-worthy is less important within these systems than the ability to perform recall functions that suit the needs of the moment. The electronic archive does not, in fact, need to leave anything out.

Archives mobilize and depend on particular recording technologies, which inform not only the means of recording but also what counts as "archivable content." The archive machine doubles as a history machine. It establishes an "archival economy" that establishes the terms for significance.[67] In recording, these technologies also invent the terms for the originality and future relevance of that which is documented. Past and anticipated events alike, then, potentially shift, both in their recordability and in their recognized relevance, through recording technologies. Histories and futures emerge—are programmed and computed—through archival machines. The electronic archive, as a recording technology, requires a certain newness, an "original proposition" as the basis for archivization.[68] But the means by which newness is arrived at shifts. Electronic archives do not consist of something initially new and then indelibly fixed in a relatively permanent medium such as print. Instead, newness emerges with each computation and transmission. Seemingly trivial data may be acquired and stored in mass. With each executable program, with each search and process function, the data becomes new again. Electronic records invent the terms for their relevance through this operation, by making claims to newness with each processing. In this respect, nothing is without significance or possible significance. MyLifeBits suggests as much: with every documentary stroke, new material emerges as possible archivable content, from number of visits to the dentist to phone calls made daily. The project is, as the researchers write, "inexhaustible." Because there is nothing that potentially would fall outside the walls of the archive, the electronic archive can continually renew everything through digital operations. Any "waste" in the record can be rescued, instantly, as an item of relevance. Through the archival program, newness repeatedly occurs, emerging through processes of execution, searching, and storing, as well as emulating, migrating and refreshing. These strategies all make it new—however trivial the content—again and again.

Electronics shift the practices of collection, archivization, and memory; they give rise to a new archival economy. The other side to this economy is, of course, the data that does not undergo searching, recall, and refreshing but, instead, sits idle in storage. Over a relatively short span of time, this saved electronic data that is not accessed begins to decay, then is lost and forgotten. If, a decade hence, researchers should have reason to use this long-neglected data, they may find it to be completely inaccessible, effectively lost from the record. In many respects, the electronic archive not only constructs but also erases events. Data is not lost because it is not archived, however; it is lost because it is archived, because it is digitized and entered into the seemingly endless electronic stores that are also increasingly volatile sites of memory. Economies of erasure, as much as economies of memory, emerge with the electronic archive. We have the capacity to store everything for possible recall, but these same extended memory technologies are capable of generating oblivion in other ways—not least of which is through the technological obsolescence that is so critical to their further development. Digital memory is volatile in more than one way. From the on and off of RAM, to the obsolescence of electronic formats, to the disposal of any analog or material version of archived material to save storage space, we forget in distinctly electronic ways.

Electronics are shot through with novelty and obsolescence and yet now operate as the most comprehensive technology ever developed for archiving. These electronic archives enable infinite capacity for storage and short-term searching, but the temporalities that they process do not extend through the generations (unless these generations shrink to the span of Moore's Law). New structures of memory imply new structures not only of forgetting but also of erasing and demattering. The constitution and deconstitution of material culture, then, occurs through these distinct mechanisms of electronic technologies.[69] With digitization (a kind of demattering), analog originals are often discarded or stored in inaccessible locations, so that there is no "original" or accessible material version to which to refer. But here is another recursive loop, where modern materials—produced through the "fad machine"—are more prone to breakdown and decay. Preserving a digital, leak-free version of these material objects becomes a way to circumvent the threat of contamination and decay. In many cases, modern objects are not meant to last. Conservators encounter the dilemma of whether they should preserve decaying materials or allow their disintegration to take place as a more accurate reflection of the objects' trajectories.[70] Digital archiving may offer an initial relief from the problem of this type of material decay, from cracking plastics and corroding metals; the matter of decay does not vanish, however, but relocates and is even amplified through electronic technologies.

"Conservation," as anthropologist Victor Buchli writes, is "anything but that: it is a very active and deliberate process of materialization; it 'conserves' nothing but 'produces' everything."[71] By delaying objects in a seemingly ideal state, conservation produces a fixed sense of material culture that allows narratives of industrial progress to persist. It is for this reason that demattering is typically excluded from these narratives: it fundamentally goes against narratives of progress.[72] Such material and even alchemical transformations consist of shaping material culture through the assumptions of longevity, permanence, and technical evolution. Yet another alchemy emerges here, the alchemy of electronics, which reveal, through their material transformations, how these technologies contribute both to distinct modes of demattering and to reconfigurations of material memory. In these museums of failure, we begin to witness the dissolution not just of electronics but also of a material culture guided by permanence and duration in contradistinction to transience.

Even in spaces beyond electronic data, the hardware of electronic objects does not directly stimulate our memory simply through its sheer material presence. Electronic objects, whether computer hardware or ancient mobile phones, become inert as physical objects as they are disconnected from any functioning system that would make their operation more intelligible. The assumed association between artifact and memory has become subject to question.[73] Furthermore, with electronic archives, no longer do we activate memory through a store of objects; instead, items must be continually called up in rapid succession, a situation where the museum or archive becomes a "flow-through and transformer station" by "unfreezing" its objects.[74] Objects are unfrozen from their distant places of storage, but often the transmission and activating of information occurs through circuits that render material in electronic formats. With these formats, objects come to seem not only less "solid" but also less permanent. In order for objects to remain as active elements within memory, they need to be activated and recalled continuously and migrated across platforms. The longevity of electronic archives depends on prolonging this condition of impermanence through the permanent act of transfer.

Electronic objects and data disappear at a regular rate. "Page Not Found" is a common message transmitted to users of the Internet. Sites on the Internet are so unstable that a (now-obsolete) project, the Museum of E-Failure, has sprung up to attempt to catalog these "ghost sites."[75] Online kitty litter warehouses and personal Web sites alike plunge into the irretrievable ether at a regular clip. The Internet may be the most thorough archive of impermanence, a fact that is made startlingly clear through projects that attempt to archive Internet materials. The Internet Archive is a project that has established an "Internet library," that allows open access to an ongoing collection of Internet materials, including "texts, audio, moving images, and software as well as archived web pages." The objective of such a collection is to "prevent the Internet—a new medium with major historical significance—and other 'born-digital' materials from disappearing into the past." The Internet Archive's efforts extend to transforming ephemera into "enduring artifacts," as well as "reviving dead links," so that when the "404 - Page Not Found" error is received, "archived versions" of these lost sites are available. In their collection, the Internet Archive also offers a "Way-Back Machine" that displays Internet sites as they looked in a particular era.[76] This project then undertakes the continual translation of obsolete electronic materials into legible format. At the same time, the Internet Archive is a project that encounters the dilemma of how to archive itself, of how to store and reproduce in electronic format its own content and electronic technologies that are also subject to rapid decay and volatility.[77]

"The permanence of the archive," media theorist Jens Schröter writes, "is changing."[78] Material stored in electronic format undergoes rapid and inevitable decay and may become inaccessible, unless it is continually migrated to new formats. As much as the computer appears to be a universal machine, it is also a universally migrating machine. It depends on an archival economy that requires continual reformatting and that promises permanence only through constant migration. In many cases, electronic material remains accessible because of its unwitting archivization and transmission through information networks.[79] Digital versions undergo data loss but persist as grainy derivations because they are transmitted and retransmitted. Clearly, this points to yet another shift in the electronic archive, where to store material away in secure vaults does little to ensure the preservation of material. Rather, electronic material only persists because it is in use, because it is transmitted and transformed, migrated across platforms, emulated and recovered from any number of obsolete storage formats, whether floppy discs or ancient hard drives. While efforts are made at "future-proofing" electronic materials by attempting to ensure their longevity through some universal format, it has become evident, more and more, that no single electronic format will offer such permanence. Instead, if they are to persist, electronic archives will have to operate according to processes of continual migration and emulation. Electronic material will have to be recovered, transformed, and retransmitted on the order of at least every ten years, if not more often. Emulation, or the process of simulating earlier operating systems and applications through computer programs, together with the migration and reproduction of material in new electronic formats, puts archival material in a state of continual transformation.

Of course, this raises questions about the extent to which each transformation effectively creates a new entity—not a copy as much as a "correspondence," as N. Katherine Hayles suggests.[80] This notion of correspondence points to the active processes of translation and emulation that occur across media types and within shifting media technologies. Such correspondence, or emulation, could even be construed as a kind of "salvage program."[81] In this salvage program, archivization occurs less through copying and more through a process of rescuing the electronic debris from the scrap heap through acts of translation.[82] Salvage is a distinctly waste-based operation. It requires sifting through and continually reevaluating the possible use and value of electronic material. But with each act of emulation, the version changes; salvage transforms, puts to use, repurposes. The process of preservation, the sense of permanence, the notion of an inalterable material culture once so central to the archive—all of these have been shot through with the transience, obsolescence, and mutability of electronic materials. Emulation, as a practice of salvaging, further allows opportunities for deviation, interference, and creative interpretation. The electronic waste of history will require continual refurbishment and reinterpretation. Perhaps now that "electronic waste" has become a carrier of our cultural and material lives, we may turn to consider how to salvage so much lost material.

Computer systems store more than the details of pottery shards, however; they also contain critical information on the operation of electronic systems all around us, from power grids to transit networks to banking systems. These same systems require a certain archivization and emulation in order to maintain their operation—and avoid catastrophic failure—in real time. Kittler points to the more pressing need to develop a strategy for electronics to archive electronics, that impossible condition that may only be possible through emulation. Without this archival strategy, everything from "early-warning missile systems" to "weather forecasts and nuclear plants" may reach terminal states of incompatibility and indecipherability.[83] The side effects of technological obsolescence do not just include piles of obsolete gadgets, overflowing archives, and decaying data. The transience that electronic technologies introduces extends to their own operation, a situation that could be fatal, when we realize that innumerable components of critical systems are constantly on the verge of obsolescence and system failure. This is a system that unwittingly undoes itself, as Ernst suggests when he writes, "We have come to the point where the world no longer experiences itself in terms of life evolving in time but rather as a network interfering with itself."[84] The real-time electronic archive does not just collect the everyday; it orders and possibly disrupts the everyday functioning of so much technology that surrounds us.

The program of technological obsolescence possibly reaches such a point of advanced failure that it even undoes itself. To address the fallout from technological obsolescence, it may be necessary to find another program—a salvage program that is capable of recovering and repurposing electronic material. This salvage program may need to operate with a memory that is attentive to waste. Waste in the archive presents the return of outdated, forgotten, and otherwise silent material. Waste is interference; it comes in the form of the obsolete, the failed and broken down. An attention to waste is an essential part of understanding electronics, those technologies programmed for obsolescence and in need of a salvage program. Such a program would exhibit an enlarged understanding of the ways in which waste is critical to the process of material transformation and revaluation, of disposal and recovery.[85] The electronic archive—of objects and data—brings renewed focus to this dual operation of disposal and recovery. Waste and the memory of waste operates in that murky space of salvage, a space that does not lead to the usual historical narratives or repeated performances of progress.[86] Instead, with the breaking and broken-down technologies, we can salvage more than technology; we can go so far as to recover the imagining that these technologies engendered. Perhaps, in the end, electronic devices, as well as electronic archives, may become sites more aligned with processes of material release and decay. This chapter attempts to salvage the overlooked dematering that takes place in the museum and archive, those principal sites for the processing of electronic material culture and electronic memory. By considering the obsolete objects, the limitations of preservation, the legacies of failure, and the forgotten technological marvels, it is possible to develop further this dynamic, alchemical quality of electronics, but only, as the next chapter explores, by turning our attention to questions of salvage and decay.

International Computers Ltd. instructional material, ca. 1970, Science Museum of London. (Courtesy of Fujitsu.)

FIVE

Media in the Dump

SALVAGE STORIES AND SPACES OF REMAINDER

He could tell at a glance that these ancient machines took up most of the storage space; they lined two entire walls, from ceiling to floor. Most of them had a layer of dust on them. The window space, too, was filled up by machines for sale, all second-hand, nothing new. Like a junk store, he thought morbidly. His experience went entirely against used merchandise; it made him feel queasy even to touch dusty, dirty-looking objects in second-hand shops. He liked things new, in sanitary cellophane packages. Imagine buying a used toothbrush, he thought to himself. Christ.

—PHILIP K. DICK, *In Milton Lumky Territory*

Salvage Stories

Having moved through the material and spatial registers of fossilized chips and screens, plastic packaging and electronic memory, this study arrives at the most obdurate, if disparate, aspect of electronic waste—that formless mass of peripherals and scrap, wires and printed circuit boards, that surfaces and settles in the dump and junkyard as the cast-off dregs of technological progress. This terminal tale then settles with rubbish, where electronics have ultimately reached the end of their operability and so collect and sediment in landfills. In these sites, there are two stories that emerge to reveal much different aspects of waste, electronic and otherwise. One story concerns a project crew of garbologists cutting core samples through landfills and sifting through rubbish to obtain a picture of contemporary consumption patterns. Bottles and burgers, ancient newspapers and mechanical relics, diapers and wrappers, all of the things we have used up are excavated deep from the steaming bowels of these recently sedimented landforms. The workers, in white jump-

Dismantling electronic waste and removing gold from circuit boards with aqua regia, Guiyu, China, 2002. (Photograph courtesy of Basel Action Network.)

suits and waders, enter this debris into a detailed inventory as evidence of our consumption activities. Here is a record of all that we coveted, possessed, and abandoned, sampled and tabulated from the formless sludge of decomposition.

The other story involves a picture of a worker suited in galoshes and rubber gloves standing under a lean-to, surrounded by muddy ground. In the worker's hand is a printed circuit board that he dips into an acid bath in order to extract tiny remnants of gold. Similar to the garbologists, this worker also sifts through the fallout of contemporary consumption, but for a much different purpose. He salvages valuable materials from electronic waste for resale because this material has been diverted from the landfills of developed countries and sent to developing countries for recycling. This diverted waste resurfaces in the scrap yards and loading docks of China and Nigeria. Waste not fit for Western dumps, due to either the lack of available landfill space or the high level of toxic substances in electronics, is, then, partially excluded from the record of consumption that the garbologists so meticulously compile.

Each of these operations is a salvage practice, a retrieval of wasted material, whether salvaging gold from discarded circuit boards or salvaging consumption data from the formless record of contemporary rubbish. Each of these salvage operations deals with the waste of contemporary culture. But the similarity between these salvage practices ends when we take into account the very different circuits of disposal in which electronic waste moves and settles. In chapter 3, I began my discussion of how electronics tip into these circuits of disposal, where the initial displacement of waste gives rise to places. In this discussion so far, I have addressed those spaces prior to and in transition to the dump, from the shipping container to the archive. Here, my intention is to dwell on the dump and those practices in and around the landfill and junkyard, including salvage and recycling. As mentioned throughout this study, electronic waste is often sent to developing countries under the guise of recycling. As the Basel Action Network indicates, up to 80 percent of electronic waste from the United States and up to 70 percent of electronic waste from Europe is shipped to developing countries.[1] Electronics may be diverted from Western landfills, but their "recycling" is often just a deferral until they reach another, if more distant, landfill.

This chapter registers the final stages of electronics in pieces and the processes of materialization that unfold as these fragmented machines scatter and travel across the globe, often far from their sites of initial consumption and use. While the spaces prior to the dump often generate multiple practices for the recuperation of value, the dump also is a space conducive to continually picking over the dregs, for rubbish is inexhaustible. Waste sticks and congeals; spaces of delay extend into spaces of indefinite remainder. As electronics sediment and begin to break down, they even create an unwitting archive of material, temporal, and ecological effects. These uncanny archives, in contrast to the more deliberate archives of the previous chapter, are the sites where the distinct salvage practices discussed here are located. This chapter dwells on these final staging grounds, where waste disposal does not give rise to absolute dissolution but, rather, provokes questions about how salvage practices deal with and transform remainders (infinitely deferred, but remainders all the same), how they recuperate value, and how they engage with the inevitability and irreversibility of waste.

"Textures of Decay"

The landfill is a kind of archive, which assembles not through deliberate or comprehensive collection but, rather, through a default accumulation of wasted matter tightly packed in airless cells. Deep within the mounds of refuse, an anaerobic environment develops, where materials are preserved unwittingly, simply through the lack of oxygen, light, and water.[2] Biodegradability in landfills undergoes a state of arrest, so that most dumps end up mummifying their contents.[3] Landfills ensure the longevity of the already extended life span of most materials. Electronics are embalmed, plastics endure, chemicals linger and spread, simultaneously. Wasted matter is preserved in this other archive, not as a collection of items for posterity, but as objects whose ecological duration far exceeds their cultural relevance. In this other accidental archive, which is far more disorderly and formless than even the most decrepit collection of computing history, it is possible to observe the transience and breakdown that characterizes waste and electronics.

The decay of waste occurs through temporal orders that span from the instant (of disposability) to a more extended geological history or earthly time. The landfill preserves this collision of temporal orders; it operates not just as a store of discarded objects but also as a record of technonatural relations that bear the imprint of shifting temporal and material conditions. Through the decay of material culture, it is possible to observe the landfill as an ecological archive. An unwitting staging ground for the breakdown and demattering of wasted materials, the landfill contains a record of contemporary consumption, the duration and toxicity of materials, and the transformation and remaindering of materials. It is a kind of "garbage museum" that at once preserves remainders but also generates new possibilities for material transformation.[4]

Debris is often one of the most telling registers through which to understand material cultures. In this sense, archaeologist Michael Shanks suggests we turn our attention to these relatively neglected "material textures of decay."[5] Beyond preservation and order, ruination is a formative and critical dynamic within material cultures, revealing how and where things fall apart and what material practices and geographies emerge to process this debris. The landfill is an ideal site in which to study such textures of decay, because when things break down, we encounter the effects and processes of materiality.[6] These effects and processes of material decay extend beyond the sheer fact of physical material breakdown, however, and encompass distinct temporalities and landscapes, as well as the practices and politics of salvage. When electronics break down and become formless, they split apart from the scripted spaces of preservation, progress narratives, and technological fascination.

Electronics further migrate across geopolitical divides to generate other salvage practices that must deal with the decay not just of technological imaginaries but also of toxic materialities. The salvage practices discussed in this chapter describe the actual repurposing of these materials. They also refer to the recovering of relations that are embedded within the final stages of handling electronic waste. From the debris and decay of electronics, it is possible to develop expanded salvage practices that turn over the imaginings, politics, economics, and geographies of electronic waste, in addition to the scraps of gold and copper that can be extracted from these machines. The fossils of leftover electronics make these relations resonate, and the natural history method enables the narration of these sedimented effects. In fact, Benjamin's salvage practices made use of archives and fossils as waste from the past that could be recycled to make available unexpected narratives—a form of ragpicking.[7] On one level, this form of salvage is striking in its difference from the garbology or electronic waste recycling previously described; on another level, Benjamin's analysis suggests expanded dimensions of salvage. Whether ragpicker, garbologist, or waste worker, each engages in transforming, picking through and digging up, sifting and reworking remainders—albeit for much different purposes and in much different ways.

To salvage is to repurpose objects, to recycle some elements and discard others, to reinforce materials and rescue parts that are momentarily resonant and that operate in some way that had yet to be imagined. Waste is the stratum of the past in the present that is often overlooked. Salvaging is an act of imagining, of eliciting stories that may have been buried in the everydayness of objects. Yet salvaging is at once a poetic and political activity; it rematerializes the sets of material relations that enabled the manufacture, consumption, and movement of goods in the first place.[8]

Working with waste is not a matter of simple recuperation. From the physical breakdown of objects, to the multiple sites across which they migrate, to the extended timescales and pollution that can be left behind, waste generates inassimilable remainders. Such remainders are often elided from waste management and sustainable development discourses, which propose that all forms of waste may eventually be broken down and recuperated into a usable, remainder-free form.[9] Electronics materialize, dematerialize, and rematerialize. In this process, they do not sustain a seamless return to (re)production. Instead, they give rise to irreversible effects and remainders: a constellation of electronic waste. Waste always returns. Even with extensive attempts to salvage, recuperate, and recycle waste, remainders surface and resurface, thereby challenging sustainable development models that hold out for the flawless reintegration of wasted materials for renewed production.

Salvage necessarily involves engaging with those temporalities of decay and processes of materialization that constitute the texture of waste. How do electronics die? Where do they go to die? How do they transform and decompose? What (and whom) do they leave behind? New salvage practices become necessary in order to address the irretrievable remainders that accompany waste. These practices can offer ways of engaging with waste that attempt not to project a future of management and integration but, rather, to address and recuperate waste in its complexity. There is a politics of salvage, a politics of remainder; but as Benjamin also reminds, there is a poetics of salvage, a poetics of remainder. The remainder of this chapter traces the material dissolution and decay of electronics, piece by piece, to the landfill.

Electronic Recovery, Electronic Remainder

As electronics break down at end of life, they enter several stages of devaluation, salvaging, recycling, reprocessing, and decay. Just as the manufacture of electronics gives rise to chemical fallout and wasted resources, so, too, the disposal of electronics creates debris. Thompson notes, in his study on "rubbish theory," that economic and physical decay are often discontinuous. Items become valueless, but their physical shells linger as "rubbish."[10] While this rubbish may at some time circulate back to a position of durable value, its valueless status may persist indefinitely. Electronics depreciate in a similar way, where a PC may be devalued from an initial value of 2,000 U.S. dollars at the time of purchase to a maximum resale of 150 U.S. dollars three years later. Even accounting for the sparse market for vintage computers, this disappearing value will typically never be recovered. There is yet another option for items in the rubbish category, and that is the possibility of salvage and recycling. While most electronics will never relocate to a position of durable value, they can be repaired or can be stripped and cleared of any materials of marginal value.

Electronics undergo many transformations in the course of decay. Repair and salvage typically precede recycling in the electronic disassembly process. Few electronics are repaired, due to the high costs of repair relative to the price of new machines. Remanufacturing does occur in some instances but is particularly dependent on whether electronics will be reissued in markets in developed or developing countries (with the latter often seen as a more viable market for refurbished machines). The process of remanufacture can actually conserve a large proportion of the labor, materials, and energy put into machines, since it repurposes machines into a similar form. Recycling, in comparison, focuses on salvaging and reforming materials into relatively raw substrates for renewed manufacture.[11] Although repair, remanufacture, and reuse are still possible strategies for working with inoperable electronics, they are typically less common salvage practices. With reuse, moreover, the age of the machine is an important factor in recovering any possible value. A new machine may fetch a price as high as 100 U.S. dollars if it can be repaired for reuse, while a machine more than 10 years old will have little to no value at all.[12]

Raw materials are salvaged from obsolete electronics, often by hand, by waste pickers working in conditions similar to those mentioned at the beginning of the chapter. The majority of salvaged materials sell for less than 1 U.S. dollar per pound.[13] As a report of the International Association of Electronics Recyclers indicates, the "commodity recovery values" from stripped electronics range between 1.50 to 2 U.S. dollars per machine. At the same time, these values are unstable, and because newer electronics contain fewer valuable metals and are now comprised of even more plastics, material prices are even lower than before.[14] The markets for salvaged goods also frequently fluctuate due to the changing relations between sites of manufacture and consumption, as well as the relatively minor contribution that recycled materials make to the overall supply of materials to manufacturing.[15] Many recyclers attempt to make up for these potentially erratic movements in value by trading in considerable volumes of scrap. Electronics returns to another economy of abundance—similar to the microchips discussed in the first chapter—where large volumes of electronic scrap are the most certain way to realize profits. At the scrap stage, disassembled electronics become important for their volumes of copper, gold, or steel. This is technology measured by the ton, a strange reversal of the apparent dematerialization that once characterized these electronics.

Waste, in this respect, becomes a kind of "ore," something held in large inventories and sourced from distinct areas.[16] The gathering of this ore is a project involving considerable labor. Materials are stripped and worked, altered and extracted, burned and soldered, fried and dipped. Much of this salvage work is carried out by waste workers in developing countries, who process materials in relatively informal and small-scale settings. The informal sector of waste work is, on the whole, not very well documented. But from Delhi, India, to Guangdong, China, many stages of transformation and "recovery" take place within the movement of electronic waste. Environmental scholars Ravi Agarwal and Kishore Wankhade, who work with Toxics Link, an organization that focuses on electronic waste, discuss how Delhi, India, has become a recycling center: "The presence of upstream markets, local entrepreneurship, and tiny-scale industries have made it a prime spot for trading recovering, reprocessing, and selling waste."[17]

While many of the salvage and recycling operations for electronic waste take place in backyards and alleys, this informal sector exists in a close relationship with the more formal and mainstream economic channels for material distribution. Electronic waste may be collected in formal and recognized routes for waste handling, but in the process of its disposal, shipping, salvage, and scrapping, it circulates into more informal economies and "gray" markets.[18] Well-established channels for importing used electronics exist in India and beyond. Electronic waste circulates from developed countries (including the United States, Europe, and parts of Asia) through transit points spanning from Dubai to Singapore, passing through as undefined scrap in order to ease the customs process in ports ranging from Delhi to Lagos.

Shipping containers stacked with obsolete electronics are routed and rerouted from transit point to port, labeled and relabeled as various forms of scrap or raw materials. The dismantling of electronics occurs as much through these infrastructures and routes as it does through the stripping of machines. Electronics are not labeled as waste but, instead, often travel through this more formless category of scrap. It is this same category of scrap that allows recyclers from developing countries to rescind responsibility for what happens to used electronics, for at this point, the electronics have transformed, magically, into little more than spare parts. Yet there are still many stages left in the dismantling, salvaging, and recycling of these machines. The salvage transformations that electronics undergo on their "route to the recycler" include the process of waste dealers first determining whether the machine is reusable and, if not, its potential price by weight. Machines then may be resold or scrapped, and if scrapped, they are separated into component parts, from monitors and memory to keyboards and motherboards, wires and casings, microchips and peripherals.[19] Here is the machine in pieces, where hard drives, CPUs, monitors, and chips are stripped and redistributed in secondhand markets. When all working components are extracted, the machines are then stripped for scrap. Copper wires are stripped from their housing, where hours of work may yield mountains of material but only a few dollars in return. Chips are methodically removed from circuit boards and drenched in acid baths to remove specks of gold. Waste pickers strip away at these machines that are not designed for disassembly, uncovering their toxic insides through equally toxic means of removal. They receive for their labor often just enough money to maintain a subsistence-level existence.

Multiple material transformations and exchanges take place in the salvaging of these discarded electronics. At every stage in the movement of electronic waste, material is extracted and repurposed. Electronics fall apart and are stripped and salvaged; but the spaces through which electronics move play a significant role in the process of that dissolution. The circulation of waste through spaces of remainder is a critical part of the material textures of electronic decay. The movement of waste and the different methods for processing waste span from collection and transport to assorted stages of disposal, which entail everything from incineration and recycling to dumping and exportation.[20] Exportation of waste is often discussed as an unviable method of waste handling, yet it is a common way in which materials are displaced. Indeed, while discussions of waste handling are often restricted to the more obvious channels of dumping and recycling, there are numerous other circuits through which rubbish moves, from reuse to salvaging. Objects that are used or used up do not necessarily issue straight for the dump. Secondhand goods, from clothing to furniture, may be repurposed in a number of ways. At the same time, these more innocuous goods move in different ways than goods that have a high level of toxicity, such as electronic waste.[21] As this mapping of the disassembly of electronic waste suggests, secondhand objects do not always circulate as benign objects capable of reuse.

Recirculation and recuperation are strategies essential to the movement of commodities such as electronics, but these processes are often opaque. They take place in informal economic sectors, in peripheral landscapes, performed by workers in developing countries. Recirculation also involves the transformation and conversion of materials. As John Frow suggests, "the conversion processes by which things pass from one state into another" is a critical area of material culture yet to be explored fully.[22] The processes of disposing of and destroying things not only lead to the conversion and transformation of materials but also potentially contribute to the mobility and circulation of those materials.[23] These processes may be more or less accelerated. But the conversion process and the spaces through which electronics move are replete with remainder.

Material disassembly and conversion does not just enable circulation, moreover. Circulation may also further contribute to the transformation of goods, particularly through a decline in value or fall in status. Once commodities such as electronics travel to developing countries, they migrate toward the rubbish category just by virtue of passing across this geopolitical and economic divide. As anthropologist Michael Taussig suggests, commodities that turn up in developing countries almost automatically acquire this sense of the outmoded. It is the circulation of these objects to developing countries that "releases" the "atmosphere" of objects, imbuing them with the quality of the "recently outmoded."[24] Objects manufactured in developing countries, as well as discarded objects from the developed countries, are left to molder as "relics of modernity."[25] Outmoded objects, together with toxins and waste, are cast off in this terrain that operates as a global landfill as much as a record for the fallout from modernity. However, through this record, the "power of ghosts embedded in the commodities created by yesteryear's technology"[26] come to light, revealing, at once, the promises initially offered by commodities as well as the remainder and resources that issue from maintaining the repetitive force of progress.

In addition to salvaging the material residues and peripheral geographies connected to electronic waste, it is also possible to salvage these more mythic remainders from obsolete commodities. Contained in outmoded objects are these obscured dimensions (of politics, economics, resources) that inevitably resurface with the death of the commodity. Waste pickers who salvage through the remains of dead electronics do not necessarily have the luxury of entertaining the wish fulfillment these devices promised; instead, in salvaging and recycling these machines, they reveal how these promised wishes fall apart. By stripping, salvaging, and recycling electronics to a condition of formlessness (only to be reformed), it is possible to see both the expanded materialities of these devices and the layers of politics, economies, and ecologies that sediment through them.

Recycling and Dumping

As already discussed, the process of salvage precedes recycling, as a way to strip machines of any operable parts and ready materials for transformation and return to the status of (relatively) raw materials. Distinct materials and components are extracted from electronics, from chips to copper and gold. Waste workers in developing countries employ hammers to smash cathode-ray tubes to extract copper; they heat circuit boards to remove chips; they soak these same boards in acid baths to remove gold; they extract motors from printers; they refill printing cartridges; they smash and chip plastic for melting and recovery; they strip and burn PVC wires to extract copper or aluminum; they separate hard disks to retrieve copper, aluminum, and magnets.[27] Recycling marks the transfer of these salvaged items back to production, where the metals, the plastics, and the working components are reintegrated into circuits of use. As discussed in chapter 3 however, even more than a return to production, recycling marks a return to wasting. While recycling appears to be a way to rid ourselves of remainder, to incorporate neatly all that is leftover, it in fact performs a deferral and inevitable return to the death of objects.

The transformation of waste to raw material through recycling is a way in which commodities become formless in order to be reformed. Recycling does not remove remainder or wastage; instead, it displaces and transforms waste.[28] The myth that waste may be recycled without remainder, instantly, into newly productive systems, presents a political and environmental dilemma. Not only does recycling rely on "economically viable markets" that, as Van Loon and Sabelis note, can actually take up recycled material for use in production; it also depends on the speculative "future profits" that will derive from "present waste."[29] The time between waste and recycling is supposed to be minimal, as though the fallout from linear growth may be recuperated in a cyclical time to feed back into that linear time. This collision of temporalities can present a key problem for recycling.[30] Within this equation, there is the problem presented by the assumed remainder-free and instant recuperation of waste, as well as the problem of the assumed remainder-free status of renewed production. In this model, the management of waste, its return to recycling, is a "displacement."[31] However, this displacement is not directed toward a space "outside" the "system" but, rather, occurs within systems, across temporalities, and even in fictional futures. As discussed earlier, remainder "directs us," even as we displace and attempt to reintegrate it. Remainder acquires a duration and delay, circulates through spaces, and undergoes material deformation and transformation, but it persists, nonetheless, in one form or another, as an ineradicable dust. Recycling, in this sense, is never complete and always generates even more waste.

While the majority of recycling takes place in the developing countries, some recycling, particularly initial salvage, takes place in developed countries. Electronics recycling facilities range in size and sophistication of operation. Some operations consist of a few workers who strip machines of particular components for reuse and then ship machines onward. Other operations shred entire machines. The latter process, considered by some to be one of the more advanced methods for dealing with electronic waste, consists of shredding everything into dust and separating these minute fragments into scrap categories based on their material composition.[32] In this process, materials are purposefully driven to a state of dust, as the ideal unit of recuperation. Dust that most closely approximates raw materials may then be shipped to manufacturing markets for reuse. But once again, the reuse of these materials depends on ongoing manufacturing and consumer demand. Without this demand, even the most advanced of recycling methods does little more than convert materials into idle raw materials. Whether recycling methods are "high-tech" ways of generating dust or consist of more dangerous methods of burning leftover electronics to render these materials to dust,[33] the spaces and material sediments bundled into electronics do not transform into waste-free futures.

The contradiction, of course, is that electronics are rendered functionless if they are contaminated with even a speck of dust during manufacture. As discussed earlier, dust threatens the functioning of these machines, yet dust returns as a definitive mark of the materiality and temporality of electronics. Indeed, as cultural historian Carolyn Steedman suggests, dust is a mark of the past and of the "imperishability of matter, through all the stages of growth and decay." Dust is a reminder that "*Nothing goes away.*"[34] Steedman goes so far to suggest that dust "*is not about Waste*" but, instead, "is about circularity, the impossibility of things disappearing, or going away, or being gone."[35] Through this study on electronic waste, however, I suggest that dust and waste are not mutually exclusive categories—that dust, far from constituting the "opposite thing to waste,"[36] actually increases our understanding of waste as a process involving transformation and remainder, not erasure through expenditure. Even within electronics, which are guided by a sense of the apparent ease of dematerialization and erasure, it is possible to observe just how persistent remainder is.

Processes of salvage, recuperation, and recycling are attempts to address this intractable remainder and where it goes. Yet electronics recycling not only creates renewed remainder and waste; it is also, as the Basel Action Network suggests, "a misleading characterization of many disparate practices—including de-manufacturing, dismantling, shredding, burning, exporting, etc.—that is mostly unregulated and often creates additional hazards itself."[37] Recycling potentially unleashes even more hazards to workers and environment, as toxic materials are used throughout the salvaging and breakdown of machines. Even with these dubious recycling methods, only a fraction of electronics actually enters the reuse, salvage, and recycling stream, with as little as 11 percent of all electronics being processed for recycling in the United States.[38] Many of these machines are divested from large institutions and corporations, which are required to recycle their equipment. But many current recycling practices are difficult to trace fully, and depending upon the methods used may generate effects that are as toxic as, if not worse than, landfilling. As the Basel Action Network indicates, the remaining electronic waste stream is sent to landfills or incinerators.[39]

The dump is a site where we encounter this fossil record in high relief. Garbologists picking through the recent remains of consumer culture or waste pickers in developing countries both work with the accelerated fossils of electronics. Sifting through these dead electronics—the sediment from compulsive upgrades—waste pickers may discover that the electronic mode of decay does not extend to rot but, rather, to leakage and contamination. These devices enjoy a plastic persistence and know nothing of biodegradability. Electronic material does not admit for total decay, even though the Long Now Foundation has established, through its "Digital Dark Ages" project, that digital media, including CDs, tapes, and files, all functionally decay typically within a matter of five years. Rates of decay may even accelerate in tropical climates, where VHS tapes have become almost completely obsolete, as the humidity creeps through magnetic plastic tape to render it inoperable. Yet, from the initial rendering of inoperability to a state of complete dust, there is a protracted process of wasting, decay, and sedimentation. This sediment develops not just through the making of goods but also through their unmaking.

In the dump, electronics cohabitate with indiscriminate landfill refuse. Whether at the end of the recycling process in developing countries or at the end of life in developed countries, electronics that do not undergo salvage and recuperation instead migrate to the dump. Electronic waste may travel the ocean as it passes into networks of recycling, but even such distribution is not enough to ensure that material will be reused. A large quantity of electronics sent for recycling in developing countries is in fact dumped instead of recycled, as the process of recycling proves to be too cumbersome or unprofitable. "In open fields, along riverbanks, ponds, wetlands, in rivers, and in irrigation ditches," the Basel Action Network documents, you will find "leaded CRT glass, burned or acid-reduced circuit boards, mixed, dirty plastics including mylar and videotape, toner cartridges, and considerable material apparently too difficult to separate."[40] These materials, together with the residues of ash and acids from electronics recycling, are scattered across landscapes in developing countries that are, in many cases, the global landfills for developed countries.

The version of dumping found in these cases is an open dump, in contrast to the sanitary landfills and incinerators of developed countries. But even in the space of the relatively impermeable landfill, now the most common method for waste disposal, heterogeneous materials mix in an equally indiscriminate way. The architecture of the landfill accretes through the sedimentation of trash, layers covered with earth and compacted into airless cells. The landfill settles, shifts, and subsides, generating methane gases and carbon dioxide. Material of any sort, whether paper or diapers, electronics or food scraps, is buried together in a space of "seemingly final disposal."[41] But this shifting architecture decomposes into the soil to expel greenhouse gases and heavy metal runoff, as well as intractable and scattered objects that refuse to decay.

Disposal may be "seemingly final," yet there are still multiple ways in which waste may be recuperated and in which remainder may resurface. Indeed, the seeming finality of the dump has been the source of inspiration for various proposals to redesign the dump as a space of storage, reuse, and flow. "Sorted dumps" have been one way to imagine organizing dumps according to materials and location, so that they may be more efficiently mined in the future.[42] A dump is, on one level, a repository of ore for possible future use. To this extent, the dump, as proposed by some, may even be obsolete, an ancient and inefficient way of dealing with abandoned materials. Mira Engler describes proposals by Dutch landscape architects to use dumps as "transit points," or as a "temporary storage space," where materials are stored for eventual recycling. Even more, these landfills may become the next mines, where instead of dismantling entire mountains for minerals, we can turn to these hills of consumption to extract materials.[43] Presented in these future visions for more ideal dumps is the persistent presence of waste as an "unwanted surplus"[44] that may at sometime become valuable again. Yet this vision relies on the persistent belief in some future ability to manage waste free of remainder: if we are not able to solve our waste or environment dilemmas today, they will no doubt become "technologically manageable" in the future.[45] Continuously present in these model future dumps is the question of remainder. Remainder is present in the form of leftover electronic scraps, as well as the irreversible effects of pollution and the damaging disparities that can emerge through the unequal economies of waste handling and dumping.

As this tour through the circuits of electronic waste further attests, the dump is a "seemingly final" space of disposal in yet another sense, as the extended effects of commodities persist well beyond burial. Even when capped under the ground, these materials belch and leach and generate pollution and methane through their decomposition. The most fluid of proposals for the reintegration and recycling of waste still generates an intractable spread and persistence of pollution. Indeed, as the Basel

Action Network indicates, "About 70% of heavy metals (including mercury and cadmium) found in landfills come from electronic discards."[46] Just as the production of electronics involves the release of numerous hazardous materials into the environment, so recycling and dumping of electronics unleashes a tide of pollutants, from lead and cadmium to mercury, brominated flame retardants, arsenic, and beryllium that spread through the soil and enter the groundwater. From manufacture to final decay, electronics seep into the aquifer and subsoil, settling into longer orders of time and more enduring chemical-material conditions.

When operable, electronics hardly seem to constitute a form of hazardous waste. Perhaps it is for this same reason that electronic waste is not always agreed on as a form of hazardous waste.[47] Yet each of the materials listed in the preceding paragraph is known to have deleterious effects on humans and environments.[48] The substances contained within electronics are precarious. They leak and spread, contaminating that which they touch. Yet another form of dematerialization, then, takes place with electronics, where the boundaries of objects break down, eroding and corroding other materials. At the same time, electronics perform another sort of rematerialization through pollution and through remainder. "Pollution surprises," writes anthropologist Marilyn Strathern, "by its untoward nature, an unlooked for return; yet those involved in the activity of waste disposal know that one cannot dispose of waste, only convert it into something else within its own life."[49] It may be possible to recycle or transform materials such as electronics into raw materials, component parts, or adaptable architectures. But in these conversion and salvage practices, it is inevitable that pollution, residue, and remainder will persist. No amount of future reintegration or reuse can negate the present effects of waste.[50]

The presence of waste and remainder suggest that we should direct our attention to the ways that things fall apart, the material textures of their decay, and what is left over. Only by turning to these processes of dematerialization, or demattering, is it possible to attend to the complex material effects of electronics. Analyzing the ways in which things—here, electronics—fall apart is critical for developing a more thorough understanding of their processes of materialization. In this respect, Buchli argues, "What is more important probably is not to study the materializations themselves but rather what was wasted towards these rapid and increasingly ephemeral materializations."[51] These processes of materialization extend to the "cultural work" that informs how objects dematerialize and transform to rubbish. Through such a study of digital rubbish, it may be possible to capture material culture more fully—not as fixed and settled, but as contingent, ephemeral, and even wasting.[52] As this chapter attempts to document, the wasting that occurs through these processes of materialization has a texture and remainder that cannot simply be erased from the material record. Electronic waste directs us toward these materializations and reminds us that irreversibility and remainder challenge the prevailing models of "waste management," which do not account for remainder. This same remainder and irreversibility create a fossil record. These fossils, the record of transience that accretes and does not reintegrate into a renewed story of technological evolution, allow us to consider "what was wasted" in these materializations.

By picking through the dump and by expanding the scope of salvage practices, it is possible to observe all that was wasted. At the same time, such a formless mass can become something other than the guilt of discards, or fodder for renewed production. The dump, instead, gives rise to new imaginings. "The dump," as architect Rem Koolhaas suggests, "has potential; it attracts scavengers."[53] This potential emerges from the apparent formlessness of the dump and its dirt, where objects become indistinct, even putrid. These objects have moved from form to formlessness, yet, as Douglas writes, "formlessness" becomes "an apt symbol of beginning and of growth as it is of decay."[54] The question Douglas poses from the rubbish is how "dirt, which is normally destructive, sometimes becomes creative."[55] But as this chapter suggests, such creativity and growth are not simple acts of reintegration and return to production and progress. Instead, the waste that surfaces here requires us to ask how remainder may "direct us" not to simplify things but, instead, to work through the complex layers and effects accreted through materializations. In this way, it may be possible to salvage not just these material relations but also the politics and poetics of matter.[56] The conclusion that follows attempts to open up the possibilities of such an encounter with remainder.

Dismantling parts—electronic waste, China, 2007. (Photograph courtesy of Greenpeace / Natalie Behring-Chisholm.)

Worker strips wires—toxics e-waste documentation, China, 2005. (Photograph courtesy of Greenpeace / Natalie Behring-Chisholm.)

Electronic waste, London, 2004. (Photograph by author.)

Conclusion

DIGITAL RUBBISH THEORY

In these reflections on the multiple, on the mix, on the speckled, variegated, tiger-striped, zebra-streaked aggregates, on the crowd, I have attempted to think a new object, multiple in space and mobile in time, unstable and fluctuating like a flame, relational.
—MICHEL SERRES, *Genesis*

If you make a motor turn in reverse, you do not break it: you build a refrigerator.
—MICHEL SERRES, *The Parasite*

Zero Waste

Two waste fantasies occupy the imagination of Kevin Lynch at the beginning of his study *Wasting Away.* These are opposing fantasies, one involving a "waste cacotopia," a society that produces waste rampantly and profligately, destroying everything it touches. The other involves a waste-free society, where there is "no more garbage, no more sewage; clean air, an unencumbered earth." In this place, "Plants and animals will be bred to reduce their useless parts: stringless beans, boneless chickens, skinless beets."[1] There would be no parasites, no weeds, no stray animals, no trash, no dirt, no dust, and "no spills, no breakage, no smoke or smog." Silence would prevail, and "friction" would be "reduced to the minimum needed to keep us erect and keep things in their place." As part of this friction-free campaign, "the edges of the continents" would even be "smoothed to reduce the tidal losses."[2] This vision of a waste-free society seems as startling as the wasteful one. As Lynch writes, "One fantasy has bred another, and neither seems attractive."[3] Yet it is typically these two polarities that are presented in relation to waste, producing it

Computer keyboards—electronic waste documentation, China, 2005. (Photograph courtesy of Greenpeace / Natalie Behring-Chisholm.)

in abundance, while simultaneously imagining the utopic possibilities of a waste-free society. Perhaps, however, the strange prospect of each of these worlds presents cause for reconsidering the intractability of waste; and by focusing on waste, it may be possible to unearth overlooked relations within the politics and poetics of things.

Strategies for dealing with waste often proceed by imagining its elimination: a society of "zero waste." In resonance with the second of the two preceding waste fantasies, zero waste is a concept and movement that has emerged as a response to the profligate wastefulness of Western societies and, in particular, to the wastefulness of manufacturing processes.[4] While the objectives of zero waste—to minimize waste in the waste stream and to develop ways of redesigning industrial processes—are important for addressing waste, "zero" may be a misleading approach to waste. Waste management and sustainable development scenarios typically consist of proposals not just to eliminate but also to make newly productive and profitable the remainders from previous cycles of production and consumption.[5] In these scenarios, the assumption is often made that if markets emulate "nature," then it may be possible to arrive at perfectly streamlined material economies. In this way, economies may also become "natural."[6] But the sense of the "natural" at work here is twofold: it is supposed, on the one hand, that the "natural" condition of environmental systems is to be at "harmony" (i.e., nature produces no waste) and, on the other hand, that material economies will ideally emulate and advance such natural harmony through the eventual progress offered by new technologies and systems.

Things wear out, fail, and break; systems of value shift and render some things worthless; transience takes hold of even the most enduring artifacts, practices, and places.[7] Rather than encounter waste, failure, and transience as conditions in need of elimination, it may be possible to consider these conditions as constitutive elements of material processes.[8] As I have argued in the pages here, there are multiple ways in which electronics generate waste. Rather than imagine the simple elimination of this waste, I have traced these residues from the fossils of manufacture to the sites of technological imagining. By working through these remainders, I have attempted to demonstrate that waste is more than a heap of defunct objects; it is also a mixture of flickering and mutable relations. Through waste, it is possible to think a "new object." This natural history of electronics, then, proposes a different sense of the "natural," which does not purify this category as an (ever-receding) ideal to move toward but, instead, considers how new natures are always in the making, emerging in that fluctuating mix of machines, nonhumans, and people. Wastes, too, are a critical part of this natural history: they are not excrescences to be weeded out at some future date. If waste, as Hawkins suggests, is "inevitable,"[9] this is not because of some tacit agreement with rampant forces of production and consumption but because no society can entirely rid itself of waste. By acknowledging the inevitability of waste, it is possible to think of it not exclusively as a menace to be eradicated but as a formative part of our material lives.

Visions of a waste-free future potentially obscure the very conditions through which waste emerges. Once waste is understood as an integral aspect of processes of materialization, it is no longer possible to imagine its complete elimination or to position it simply as raw material to be fed into friction-free futures. Instead, the persistence of waste occurs in part through the unavoidable remainders that do not easily recycle into new systems of production or that are left behind as the pollution and residue from previous activities. Waste does not consist just of the fossils from past cycles of production and consumption; it is also the remainders generated from continually unanticipated futures. When proposals are made for a "solution" to the waste "problem," waste is often displaced back into the same productive mechanism that produced waste in the first place.[10] But as discussed in chapter 5, such a "discount on the future," as Van Loon and Sabelis characterize it,[11] does not account for the "costs of irreversibility,"[12] which will contribute to future complexities beyond our present methods of accounting. By appending "zero" to waste, we obstruct the possibility of considering how irreversibility and remainder emerge as integral aspects of waste.[13]

As long as our basic approach to waste depends on its eventual and continual eradication, it will be difficult to grasp the ways in which waste emerges and operates—as generative and dynamic and, as Hawkins suggests, as the "terrain of ethics."[14] Arguably, the development of apparent waste-eliminating strategies such as recycling not only obscures the inevitability of waste[15] but also defers the ethical aspects of how we attend to waste—whether we bury it, ship it to developing countries, or leave it to future generations to trawl through. It may be possible to move beyond a "dos and don'ts" approach to waste, as Van Loon and Sabelis write, and instead "to generate a radical reconceptualization of waste itself."[16] Rather than consider recycling as the instant reintegration of waste into the market, it may be possible to attend to the ways in which waste—as a mutable and relational object—offers "possibilities for the unexpected, the creative and the ethical."[17] The creative and ethical aspects of waste are often typically elided, particularly in campaigns for its elimination or reintegration, yet it is from these remainders and fragments that it is possible to realize the political and poetic registers of matter. Remainders direct us not toward the recovery of "wholeness" but toward new possibilities for working with the "scatter" of the world. Waste allows the possibility for "imagining a new materialism," as Hawkins suggests, resonating with the material imaginings put forward by Benjamin.[18] But the question of how this materialism emerges and registers still persists.

Garbage Imaginaries

In many cases, attempts to imagine a new materialism for electronics extend from improving the life-cycle impacts of these devices, minimizing their ecological footprint, improving working conditions for fab workers, and banning the exportation of wastes to developing countries for "recycling."[19] In addition to stricter environmental policies and regulations, design is often seen as a key way in which to improve the environmental impact of electronics. Numerous design projects address ways in which to eliminate, reincorporate, or otherwise track remainder, from point of manufacture on to consumption and disposal. These projects, often based on life-cycle analyses, suggest that waste may be minimized by altering design approaches. This is an ideal way in which to "regulate" waste, as Molotch suggests, because "design determines about 80–90 percent of an artifact's life-cycle economic and ecological costs, in an almost irreversible way."[20] Hazardous materials and landfilling can be avoided through the more careful design of electronics. In this way, Greenpeace's "Guide to Greener Electronics" suggests that electronics companies develop "a chemicals policy based on the Precautionary Principle" and phase out known hazardous materials that are used in machines, including brominated flame retardants and other "problematic substances."[21] A complex composite of plastics is also used in electronics, plastics that are difficult to reuse or recycle at end of life and that could be simplified for this purpose. If electronics companies were responsible both for what goes into machines and for their eventual take-back and recycling, then they might possibly begin to find it effective to make these devices less toxic at the outset.

Without a doubt, the reduction of hazardous materials and introduction of methods of recycling and disassembly are necessary developments within the world of electronics.[22] Within this area, there are so many projects underway that it is tempting to make a modest proposal and public appeal for someone to write a "handbook" about green machines—the sort of handbook that could be circulated to enable new ways of thinking about electronic design and production.[23] "Green technology" is not only seen as a major area of invention; it is also a complex and interesting terrain for new design projects. In an industry that is preoccupied with continual invention—where pronouncements are made about the "convergence" of technologies, about pervasive computing, about Web 2.0 and the death of the Internet and the end of Moore's Law—it seems appropriate to consider how that invention can extend into this other terrain.

Emerging proposals for "green electronics" or "green ICT" (information and communication technology) include schemes that address the material composition and manufacture of electronics, from computer keyboards made out of carrot and spinach extracts to mobile phones that "plant" sunflower seeds when they decompose.[24] Microchips that are oxidized through ultraviolet radiation, rather than energy-intensive furnaces, are now in prototype stage; PCs are available in die-cut cardboard, rather than a composite of plastics; and mobile phone prototypes "self-recycle" by popping apart when heated, for ease of disassembly and recycling.[25] An extensive number of electronic design projects also focus on ways of improving energy consumption within the operation of devices.[26]

Other projects document or propose interventions within the life cycles of electronic devices.[27] Some designers have gone so far as to suggest that design not only should alter at the manufacturing phase but should also extend into "everything that happens after that." In this sense, designer Ed van Hinte writes, goods should not be "impenetrable boxes" but, rather, should have "a career plan."[28] In this scenario, design extends to consumer use, commodity alterations, and eventual dismantling. Other projects draw attention to the expanded circuits and possibilities of things beyond the manufacture stage by using electronics to track trash, so that electronic devices may even become the means for possible infrastructures of reuse.[29] These tracing and tracking projects pay particular attention to the object—electronic or otherwise—as it cycles from manufacture to use and death.

Still other projects reconsider the relatively functional role of electronics in our lives and draw out the more imaginative and uncanny dimensions of these devices.[30] Repurposing obsolete electronics through reverse engineering and hacking has been one strategy not only for unpicking the assumed functionality of these devices but also for extending the practices of reuse and recycling beyond the simply material toward new technological deployments.[31] Concepts of "reuse," "appropriation," and "maintenance" are emerging as practices for investigating the possibilities of sustainable computing.[32] Electronic capabilities may, at the same time, enable other modes of encounter with environments, and much of the literature on "sustainable HCI" (human-computer interaction) has dealt with not just issues of green machines but also ways in which social networking, citizen science, and ecological monitoring and information may persuade and raise awareness about environmental issues.[33]

Together, these projects address everything from materials and manufacture to systems and new imaginaries for the use and abuse of electronics. It is a significant step toward a more "green" and creative approach to electronics. Yet the question that remains within such initiatives is whether attention to waste, as well as the extended political and economic effects of electronics, will provoke us to think about technologies differently. Designs for green electronics may be most successful when they consider not only the material effects but also the extended social, political, and imaginative terrain of electronics. This means that it may be possible to do more than just alter electronics to contain fewer contaminants, have an ease of disassembly, and be more readily reusable; we may also reconsider how electronics materialize and rematerialize across multiple spaces and practices. This natural history of electronics, then, raises questions about how to go beyond the gadget as it passes through its life cycle. Such a conception of electronic technologies potentially settles on one dimension of the life and death of these devices. However, a complex circuit of places and politics, materials and ecologies, and uses and manufacture makes possible and sediments into electronics and electronic wastes. As a thing and technology, electronics and electronic wastes are the sites of stories that exceed product life cycle and that ultimately connect up lives, labor, and imaginaries.[34]

The natural history of electronics developed here draws on these proposals and suggests that one way to develop "sustainable" electronics would be to address the multiple materialities, politics, ecologies, economies, and imaginings that give rise to electronics.[35] These technologies are not only a part of natural-cultural arrangements; they also provide insight into the ecologies we inhabit. In this sense, there are opportunities to engage with the creative and ethical aspects of electronics and electronic waste not just through improving electronics manufacture but also through linking up ecologies—political and otherwise. Supplying ICT for the developing world is just one way in which electronics can be deployed not so much for another round of consumption but, instead, to connect up communities who may not otherwise have access to electronic communications and to make these technologies less toxic in the process. Soenke Zehle suggests we revisit earlier proposals for an "environmentalism for the net."[36] Such an environmentalism might consist of "info-political initiatives" that encompass not just the digital commons but also the "broader agenda of economic and environmental justice."[37] In this way, applications are being developed through original uses of renewable energy—wireless that runs on wind power—that begin to take up a digitally relevant environmentalism that expands beyond but also encompasses less deleterious and resource-intensive manufacture and energy processes.[38]

Some of the most compelling projects to be found working in this area establish creative ways to make the environmental, social and environmental relations that emerge through electronics a site of reinvention and provocation. The "Zero Dollar Laptop" project, a collaboration by Furtherfield, Access Space, and St. Mungo's charity for the homeless, offers a series of recycling workshops that engage with obsolete electronics. The participants engaged with the project recycle outdated laptops, and install Free and Open Source Software on the machines to enable the use and creation of media files, and to provide access to the Internet. Obsolete hardware and software offer up a set of new resources, as this project demonstrates, if the terms of use shift to engage with alternative economies and exchanges. In a different approach, Graham Harwood and Matsuko Yokokoji have made the material and energy requirements of computers evident in their "Coal Fired Computers" project, which demonstrates how central coal power is to the manufacturing and firing of computer circuits, since coal still provides a considerable amount of power to our modern energy economies.[39]

The focus in this study has been to unpack the black box of electronics by charting stories that converge in the saturated soil of Silicon Valley, in the run of numbers that flicker across NASDAQ interfaces, in the global trawl of waste shipments, in the defunct machines gathering archival dust, and in the thick sludge of the landfill. In considering these places and stories, where the debris of electronics collects, I suggest there are other ways of thinking about material culture through these remainders.

Electronics constitute "materializing and transformative processes." Such processes, as Buchli writes, give rise to "new kinds of bodies, forms of 'nature' and political subjects."[40] The processes whereby materials congeal and fall apart are essential for understanding things as matter. The ways in which electronics stabilize and destabilize are bound up with technological trajectories and markets, methods of manufacture and consumption, and imaginaries and temporalities.

There is even potential in this space of imagining to consider the fantastic qualities of electronics and for a material imagination that surpasses the strictly instrumental and the progressive.[41] Remaining in cast-off objects is that same "wishful" element that Benjamin saw as most potent at the moment of their introduction. The fossils in his natural history were not without fascination; in fact, they depended on it. Without a doubt, there are many approaches to electronics that may begin to find the advantages of operating in these fields, beyond the appeal of novelty and functionality and toward a kind of garbage imaginary. So perhaps what we need are electronics that exploit and expand on the cracks, the failures, and the garbage, as a way to move toward the creative and ethical aspects of electronics and electronic waste, as a way to imagine new material relations.

This garbage imaginary is a fitting place to conclude this study into electronics and waste. The "cultural imaginary" of garbage, as Shanks, Platt, and Rathje write, "is at the heart of the composition and decomposition of modernity and modernism."[42] A garbage imaginary might emerge not just by seeing the matter of things, the fields through which they circulate, and their modes of transformation and animation; it might also emerge, as Lynch suggests, by "wasting well." If waste is inevitable, then it may be possible to begin to address how matter transforms and to draw out the moments and movements where energies, resources, values, temporalities, and spaces shift. In dirt, there is potential. Dirt rituals have existed for quite some time. To this extent, Lynch even considers the fascination of "collision derbies and the art of piano-smashing."[43] It may be possible that we need more and better ways of encountering the ways in which things run down and wear out. With a less exclusionary sense of waste, it might be possible to see that matter moves in "gradations" and, thereby, to devise "ceremonies of transformation."[44] By registering the ways in which materials transform—the processes of materialization through which things sediment—it is possible to take greater responsibility for our material lives.

But in these moments of transformation, the smashing ceremony that resonates the clearest is the one described by Benjamin in his "Theses on the Philosophy of History." He describes how, "on the first evening" of the "July revolution," the clock towers in Paris "were being fired on simultaneously and independently from several places."[45] In this moment, time no longer progressed along a chronology. With the clock towers shot out, the empty space of progressive time was stopped in its tracks. In the absence of progressive time, a shift in the *experience* of time could emerge. In the "now" of suspended progressive time, the "new" could materialize through other temporalities, not as a space of transition or even revolution, but as a space of material relation and imagination. This is a transformation that takes place not simply in succession but through a generative and waste-based imaginary that involves the politics as much as the poetics of materials. That imaginary, as described here, settles into a natural history of electronics.

International Computers Ltd. instructional material, ca. 1970, Science Museum of London. (Courtesy of Fujitsu.)

Notes

Preface

Bruce Sterling, "The Dead Media Project: A Modest Proposal and a Public Appeal," http://www.deadmedia.org/modest-proposal.html, accessed March 4, 2008. The "Dead Media" project is, at the time of this writing, now off-line. True to the topic of this book, Web sites and URLs are notoriously short-lived, and constitute yet another form of fleeting electronic media.

1. Opening the "black box" of technologies is a common practice and turn of phrase in science and technology studies to describe the making evident of technological processes that might otherwise go unnoticed. But this phrase also has a longer history of usage, and is a frequent figure within computing and information theory discourses. This study employs both senses of the term throughout. For more on this (double) use of the black box, see Michel Serres, *The Parasite,* trans. Lawrence R. Scher (Baltimore: The Johns Hopkins University Press, 1982), passim.

Introduction

Walter Benjamin, *The Arcades Project,* trans. Howard Eiland and Kevin McLaughlin (Cambridge: Harvard University Press, Belknap, 1999), 390.

Donna Haraway, "The Promises of Monsters: A Regenerative Politics for Inappropriate/d Others," in *The Haraway Reader* (New York: Routledge, 2004), 116 n. 14.

1. The Superfund designation, which dates to 1980, was initiated by the U.S. federal government to clean up hazardous waste sites with particularly high levels of contamination. A description and map of Superfund sites designated for cleanup, known as the National Priorities List, can be found on the EPA's Web site, at http://www.epa.gov/superfund/sites/npl.

2. See the listings for Santa Clara County in the EPA's National Priorities List, http://yosemite.epa.gov/r9/sfund/r9sfdocw.nsf/WSOState!OpenView&Start=1&Count=1000&Expand=2.29#2.29.

3. *Ecologies* is a term that has circulated within media studies since at least McLuhan onward to describe a focus on media environments (at times in contrast to a focus on media content). This term has been recuperated and revised by multiple contemporary researchers, including Matthew Fuller, *Media Ecologies: Materialist Energies in Art and Technoculture* (Cambridge, MA: MIT Press, 2005). The term *ecologies* is used in this study to refer both to media environments and to those more natural-cultural ecologies that occur through soil and chemicals, water, and air.

4. Gopal Krishna, "E-Waste: Computers and Toxicity in India," in *Sarai 3: Shaping Technologies* (Delhi: Sarai, 2003).

5. International Association of Electronics Recyclers, *IAER Electronics Recycling Industry Report* (Albany, NY, 2003; revised, 2004), 7. These quantities represent only consumer electronics. Commercial and industrial sectors are such large generators of electronic waste that they have until recently "driven" the electronics recycling industry, due to the electronics they discard in such appreciable volumes. Beginning in January 2009, the International Association of Electronics Recyclers was acquired by the Institute of Scrap Recycling Industries (ISRI), which produces regular newsletters on electronics recycling issues. See http://www.isri.org.

6. This figure is cited as originating with the Institute for Local Self-Reliance and is documented by the EPA. See Environmental Protection Agency, "Waste Wise Update: Electronics Reuse and Recycling," EPA 530-N-00-007, October 2000, http://www.epa.gov/epawaste/partnerships/wastewise/pubs/wwupda14.pdf.

7. International Association of Electronics Recyclers, *IAER Electronics Recycling Industry Report*, 25.

8. The European designation for electronic waste is WEEE (waste from electrical and electronic equipment). The WEEE Directive, which lists an extensive number of products that constitute electrical or electronic waste upon expiration, is accompanied by the Restriction of the Use of Certain Hazardous Substances Directive, or RoHS, which attempts to limit the inclusion of hazardous substances in electronics. For the original directives (which have been amended several times to accommodate exemptions), see "Directive 2002/96/EC of the European Parliament and of the Council of 27 January 2003 on Waste Electrical and Electronic Equipment (WEEE)," *Official Journal of the European Union*, February 13, 2003, L37/24–L37/38; "Directive 2002/95/EC of the European Parliament and of the Council of 27 January 2003 on the Restriction of the Use of Certain Hazardous Substances in Electrical and Electronic Equipment," *Official Journal of the European Union*, February 13, 2003, L37/19–L37/23. The U.S.-based International Association of Electronics Recyclers includes a similar list of electronic waste in its reports but expands the range of electronics to commercial systems and financial systems, security systems and cameras, entertainment systems, office equipment, industrial electronics (including telecommunications equipment, control systems, manufacturing equipment, and commercial appliances), and consumer electronics (including audio, video, and communication devices). See International Association of Electronics Recyclers, *IAER Electronics Recycling Industry Report*, 2–3.

9. The Silicon Valley Toxics Coalition maps out the "Electronics Lifecycle" on its Web site, where it notes that "electronics are a complicated assembly of more than 1,000 materials, many of which are highly toxic, such as chlorinated and brominated substances, toxic gases, toxic metals, photo-active and biologically active materials, acids, plastics, and plastic additives." See http://www.svtc.org/site/PageServer?pagename=svtc_lifecycle_analysis.

10. See Jan Mazurek, *Making Microchips: Policy, Globalization, and Economic Restructuring in the Semiconductor Industry* (Cambridge, MA: MIT Press, 1999). For details about the effects of specific chemicals used in the electronics manufacturing process, see the EPA Toxic Release Inventory at http://www.epa.gov/TRI/.

11. For examples of these informative studies, see Basel Action Network and Silicon Valley Toxics Coalition, *Exporting Harm: The High-Tech Trashing of Asia* (Seattle

and San Jose, 2002), as well as BAN's newer study into electronic waste in Lagos, Nigeria, *The Digital Dump: Exporting Re-use and Abuse to Africa* (Seattle: Basel Action Network, 2005). For book-length studies on electronic waste, see Elizabeth Grossman, *High Tech Trash: Digital Devices, Hidden Toxics, and Human Health* (Washington, DC: Island, 2006); Ted Smith, David A. Sonnenfeld, and David Naguib Pellow, eds., *Challenging the Chip: Labor Rights and Environmental Justice in the Global Electronics Industry* (Philadelphia: Temple University Press, 2006).

12. I use the term *discourse* as part of a material constellation and along the lines that Judith Butler describes when she argues, "To claim that discourse is formative is not to claim that it originates, causes, or exhaustively composes that which it concedes; rather, it is to claim that there is no reference to a pure body which is not at the same time a further formation of that body." In this sense, discourse is "performative" and part of the process whereby materials, bodies, and technologies unfold and circulate in the world. See Judith Butler, *Bodies That Matter: On the Discursive Limits of "Sex"* (London: Routledge, 1993), 10–11.

13. Bill Brown, "Thing Theory," *Critical Inquiry* 28, no. 1, Things (Autumn 2001): 10; Mark Hansen, *Embodying Technesis: Technology Beyond Writing* (Ann Arbor: University of Michigan Press, 2000), 43.

14. For more information on the importance of considering how waste emerges not simply at the waste end of products but throughout the production and consumption cycle, see Kenneth Geiser, *Materials Matter: Towards a Sustainable Materials Policy* (Cambridge, MA: MIT Press, 2001).

15. Such heterogeneity of relations also describes what sociologist Mike Michael refers to as the "range of possible trajectories for the uses and 'mis-uses' (or rather, misbehavior) of technological artifacts" (*Reconnecting Culture, Technology, and Nature: From Society to Heterogeneity* [London: Routledge, 2000], 10).

16. N. Katherine Hayles has convincingly argued this point, and this study is informed by her research. See N. Katherine Hayles, *How We Became Posthuman: Virtual Bodies in Cybernetics, Literature, and Informatics* (Chicago: University of Chicago Press, 1999).

17. The term *sediment* draws on ideas from both Benjamin and Judith Butler (as discussed later in this introduction). For Benjamin, criticism was a project that should operate through the congealing of facts: "Criticism means the mortification of the works. By their very essence these works confirm this more readily than any others. Mortification of the works: not then—as the romantics have it—awakening of the consciousness in living works, but the settlement of knowledge in dead ones." See Walter Benjamin, *The Origin of German Tragic Drama*, trans. John Osborne (London: NLB, 1977), 182. For Butler, sedimentation describes temporal and material practices that inform material effects (of power) as they concretize and transform. See Butler, *Bodies That Matter*, 250.

18. Butler, *Bodies That Matter*, 9–10.

19. Benjamin discusses his concepts of "natural history" and the outmoded in several key locations, including, of primary significance for the purposes of this study, *The Arcades Project*. See also Benjamin's *The Origin of German Tragic Drama* and his "Surrealism: The Last Snapshot of the European Intelligentsia," in *Reflections*, ed. Peter Demetz, trans. Edmund Jephcott (New York: Harcourt Brace Jovanovich, Schocken Books, 1986), 177–92.

20. Benjamin, *Arcades Project*, 203.

21. Discussing Benjamin's "unorthodox" use of natural history, cultural and literary theorist Beatrice Hanssen writes, "By suggesting that history and nature were 'commensurable' in the moment of transience that befell both, Benjamin's study in fact contested the idealistic dichotomy between history and necessity, human freedom and nature, which it replaced with 'natural history'" (*Walter Benjamin's Other History: Of Stones, Animals, Human Beings, and Angels* [Berkeley: University of California Press, 1998], 9). A number of studies and literary texts take up and extend Benjamin's notion of natural history. Some of the more in-depth texts include Susan Buck-Morss, *The Dialectics of Seeing* (Cambridge, MA: MIT Press, 1989); Michael Taussig, *Mimesis and Alterity: A Particular History of the Senses* (Routledge, 1993); and W. G. Sebald, *The Rings of Saturn*, trans. Michael Hulse (London: Vintage, 1998).

22. Michel Foucault, *The Order of Things: An Archaeology of the Human Sciences* (1970; repr., London: Routledge, 1994).

23. Ibid.

24. As paleontologist Stephen Jay Gould argues, there have been multiple versions of natural history and as many ways of interpreting remains to build up a view of longer and larger earth processes. So while a Victorian view emphasizes natural history as a record of progress, with humans in the ascendancy, Gould, in his more contemporary rendering, takes issue with these conflations of natural history (and Darwinian evolution) with progress. If natural history has shown us anything, Gould argues, it is that life on earth proceeds through random, coincidental and contingent processes of survival. See Stephen Jay Gould, *Wonderful Life: The Burgess Shale and the Nature of History* (1990; repr., London: Vintage, 2000); Charles Darwin, *On the Origin of the Species* (London: John Murray, 1859).

25. Failure is often discussed within studies of technologies and economies as a necessary aspect of further development. See, for example, Thomas S. Kuhn, *The Structure of Scientific Revolutions* (1962; repr., Chicago: University of Chicago Press, 1996). However, I suggest here and elsewhere that failure does more than simply propel further technological developments. See Jennifer Gabrys, "Machines Fall Apart: Failure in Art and Technology," in *Leonardo Electronic Almanac* 13, no. 4 (April 2005), http://www.leoalmanac.org/journal/Vol_13/lea_v13_n04.txt.

26. Benjamin focused specifically on the outmoded, as it shattered, according to Buck-Morss, "the myth of automatic historical progress." Such a move was important because "when newness became a fetish, history itself became a manifestation of the commodity form." See Buck-Morss, *Dialectics of Seeing*, 79–83, 93. To challenge this fetishized and always-new reversioning of history, Benjamin conceived of the "angel of history" based on a painting by Paul Klee entitled *Angelus Novus*. Benjamin's angel is driven by the "storm of progress" into the future, to which "his back is turned." Instead of witnessing the future, the angel witnesses the "wreckage" that accumulates from this storm. See Benjamin, "Theses on the Philosophy of History," in *Illuminations*, ed. Hannah Arendt, trans. Harry Zohn (New York: Harcourt Brace Jovanovich, Schocken Books, 1969), 257–58.

27. There are other examples of digital or informational "natural histories," which, in one way, discuss just how "natural" or evolutionary electronics may be (as progressive technologies) or, in a much different way, suggest relations between new media and corporeality, digital code and biology. For examples, see Paul Levinson, *The Soft Edge: A Natural History and Future of the Information Revolution* (New York: Routledge, 1997); Anna Munster, *Materializing New Media: Embodiment in Media*

Aesthetics (Hanover, NH: Dartmouth College Press, 2006). The present study works with a more unruly set of material effects to constitute its natural history method.

28. For more on second nature in relation to natural history, see Theodor W. Adorno, "Natural History," in *Negative Dialectics*, trans. E. B. Ashton (1966; repr; London: Routledge, 1990), 354–58. The term *second nature* was initially deployed by Georg Lukács in *The Theory of the Novel* (Cambridge, MA: MIT Press, 1974). In this process, nature becomes culture through manufacture, and the manufactured objects that surround us begin to seem as naturalized as nature itself: a second nature. Yet this rendering of "second nature" as a process of transformation, of culture acting on nature, does not rest so easily in this natural history, which resists a "first nature" on which culture might operate. Furthermore, once commodities have exited the spaces of cultural production and consumption, these fossils constitute another nature-cultural coupling again, bound by temporal forces.

29. Donna Haraway, *Modest_Witness@Second_Millenium: FemaleMan©_Meets _OncoMouse™* (New York: Routledge, 1997), 142–43. In a related way, Madeline Akrich notes, "Machines and devices are obviously composite, heterogeneous, and physically localized. Although they point to an end, a use for which they have been conceived, they also form part of a long chain of people, products, tools, machines, money, and so forth." These heterogeneous assemblages, as identified here, can even "generate and 'naturalize' new forms and orders of causality and, indeed, new forms of knowledge about the world." See Madeline Akrich, "The De-Scription of Technical Objects," in *Shaping Technology/Building Society: Studies in Sociotechnical Change*, ed. Wiebe E. Bijker and John Law (Cambridge, MA: MIT Press, 1992), 205–7.

30. Haraway, *Modest_Witness@Second_Millennium*, 142–43.

31. For Benjamin, the word *scatter* describes not just qualities of present experience but also a method for working through the past—as an iterative and resonant practice. He writes of the historian who works in this way, "Above all, he must not be afraid to return again and again to the same matter; to scatter it as one scatters earth, to turn it over as one turns over soil. For the 'matter itself' is no more than the strata which yield their long-sought secrets only to the most meticulous investigation. That is to say, they yield those images that, severed from all earlier associations, reside as treasures in the sober rooms of our later insights" (Walter Benjamin, "Excavation and Memory," in *Selected Writings*, vol. 2, part 2, *1931–1934*, ed. Michael W. Jennings, Howard Eiland, and Gary Smith, trans. Rodney Livingstone et al. [Cambridge: Harvard University Press, Belknap, 1999], 576).

32. Theodor Adorno, "A Portrait of Walter Benjamin," in *Prisms*, trans. Samuel Weber and Shierry Weber (Cambridge, MA: MIT Press, 1967), 227–41.

33. Benjamin, *Arcades Project*, 392.

34. Butler writes, "For the concept of nature has a history, and the figuring of nature as the blank and lifeless page, as that which is, as it were, always already dead, is decidedly modern, linked perhaps to the emergence of the technological means of domination" (*Bodies That Matter*, 4).

35. Hanssen, *Walter Benjamin's Other History*, 16.

36. Sarah Franklin, Celia Lury, and Jackie Stacey, *Global Nature, Global Culture* (London: Sage, 2000), 59. In taking up this "set of debates about changing definitions of nature, culture and global" (ibid., 5), cultural theorists Franklin, Lury, and Stacey draw on Martin Rudwick's analysis to establish just how frequently interpretations of fossils have given rise to shifting definitions of nature, culture, history, and the

global. See Martin Rudwick, *The Meaning of Fossils: Episodes in the History of Paleontology* (Chicago: University of Chicago Press, 1976).

37. Benjamin, *Arcades Project*, 864.

38. Butler, *Bodies That Matter*, 10.

39. Ibid., 9–10.

40. Charles Acland, ed., *Residual Media* (Minneapolis: University of Minnesota Press, 2007), xxi.

41. Siegfried Zielinski, *Deep Time of the Media: Toward an Archaeology of Hearing and Seeing by Technical Means*, trans Gloria Custance (Cambridge, MA: MIT Press, 2006), 2–7. While this study draws on research in media archaeology, which shares many of the same interests in obsolete media and materiality, it opts instead to work through a natural history method in order to establish the ways in which things break down and do not cohere as singular media artifacts. In their breaking down, electronics generate further relations and political ecologies. This approach is less about media objects and mediation, and more about the transformative material ecologies of electronic media.

42. Marshall McLuhan and Quentin Fiore, *The Medium Is the Massage: An Inventory of Effects* (New York: Bantam, 1967), 26.

43. Friedrich A. Kittler, *Discourse Networks, 1800/1900*, trans. Michael Metteer and Chris Cullens (Stanford: Stanford University Press, 1992).

44. Eva Horn, "There Are No Media," *Grey Room* 29 (Winter 2008): 6–13.

45. Numerous studies have now employed an actor-network theory (ANT) approach to issues of science and technology. While this research is partially informed by literature in science and technology studies, particularly through the work of Donna Haraway, it intentionally does not adopt an ANT approach to electronic waste. Networks, I suggest in this study, can be one way to understand better how the distinct materialities of electronics are distributed and how they perform. Networks in this sense are not deployed as an organizing conceptual device for studying human and nonhuman relations, but rather as a term specifically situated within the material cultures of computing. This approach is further inspired by Michel Serres's use of "exchanges" and "quasi objects" to describe such transformative aspects of materiality. See Bruno Latour, *Reassembling the Social: An Introduction to Actor-Network-Theory* (Oxford University Press, 2007); Bruno Latour, *Aramis, or the Love of Technology*, trans. Catherine Porter (Cambridge: Harvard University Press, 1996); Michel Serres with Bruno Latour, *Conversations on Science, Culture, and Time*, trans. Roxanne Lapidus (Ann Arbor: University of Michigan Press, 1995); Kevin Hetherington and John Law, "After Networks," *Environment and Planning D: Society and Space* 18, no. 2 (2000): 127–32.

46. I use the term *resonance* here in the sense that Haraway discuses in relation to "situated knowledges," which she suggests means "a way to get at the multiple modes of embedding that are about both place and space in the manner in which geographers draw that distinction. Another way of putting it is when I discuss feminist accountability within the context of scientific objectivity as requiring a knowledge tuned to resonance, not dichotomy" (Donna Haraway, *How Like a Leaf: An Interview with Thyrza Nichols Goodeve* [New York: Routledge, 2000], 71). See also Donna Haraway, "Situated Knowledges: The Science Question in Feminism and the Privilege of Partial Perspective," *Feminist Studies* 14, no. 3 (1988): 575–99; Marilyn Strathern, *Partial Connections* (Savage, MD: Rowman and Littlefield, 1991).

47. I do not approach consumption and the "user" through everyday practice or mutual formation of users and technologies—areas that already have extensive and informative literatures. Rather, I understand "use" in several arguably undertheorized ways that focus on what comes after a more privileged idea of use as agency, often in the form of content manipulation. Following on a number of waste literatures discussed in more detail in chapter 3, I focus on use as *using up*—through disposal, through the labor of breaking up machines, and through the political and economic using up of geographies as global dumping grounds. For examples of "use" literature, see Bijker and Law, *Shaping Technology/Building Society;* Nelly Oudshoorn and Trevor Pinch, eds., *How Users Matter: The Co-Construction of Users and Technologies* (Cambridge, MA: MIT Press, 2003); Susan Leigh Star, ed., *The Cultures of Computing* (Oxford: Blackwell, 1995); Lucy Suchmann, *Plans and Situated Actions: The Problem of Human-Machine Communication,* 2nd ed. (Cambridge: Cambridge University Press, 1987).

48. See Daniel Miller, ed., *Materiality* (Durham: Duke University Press, 2005); John Law and Annemarie Mol, "Notes on Materiality and Sociality," *Sociological Review* 43 (1995): 274–94; Michelle Murphy, *Sick Building Syndrome and the Problem of Uncertainty: Environmental Politics, Technoscience, and Women Workers* (Durham: Duke University Press, 2006).

49. While this is not a life-cycle analysis, this research does draw on studies that analyze electronics through their material inputs and environmental costs. For examples of this approach, see Ruediger Kuehr and Eric Williams, eds., *Computers and the Environment: Understanding and Managing Their Impacts* (Dordrecht: Kluwer Academic, 2003).

50. Hansen, *Embodying Technesis,* 60.

51. Hirokazu Miyazaki elaborates on the ways in which different theoretical practices identify an "object" and its "materiality" also count as material processes and constitute an important site for material analysis. See "The Materiality of Finance Theory," in Miller, ed., *Materiality,* 165–81.

52. The use of the term *rematerializing* in this study refers to the registering of complex material processes that support otherwise apparently material-free electronics. Rematerializing in this sense is distinct from those more geographical debates concerned with whether rematerializing constitutes a return to physical matter (as opposed to cultural or conceptual concerns). See Ben Anderson and John Wylie, "On Geography and Materiality," *Environment and Planning A,* 41, no. 2 (2009): 318–35.

53. Haraway, "Promises of Monsters," 109.

54. Haraway, *Modest_Witness@Second_Millenium,* 43–45.

55. Benjamin, *Arcades Project,* 460.

56. Ibid., 459.

57. Buck-Morss, *Dialectics of Seeing,* 65.

58. Benjamin, *Arcades Project,* 871. This notion of the past sedimenting into space can also be found in Benjamin's earlier, related study of allegory and ruins, where he writes that "chronological movement is grasped and analyzed in a spatial image." He also writes, "In the ruin history has physically merged into the setting. And in this guise history does not assume the form of the process of an eternal life so much as that of irresistible decay. Allegory thereby declares itself to be beyond beauty. Allegories are, in the realm of thoughts, what ruins are in the realm of things" (*Origin*

of German Tragic Drama, 177–78). The meeting of time and space in the ruin has influenced many scholars, including Kathleen Stewart, who develops a distinct method for writing about the cultural and poetic aspects of ruins. See Kathleen Stewart, *A Space on the Side of the Road* (Princeton: Princeton University Press, 1996), 96.

59. While mining sites are clearly another waste zone related to electronics, this study does not elaborate on these sites, not only due to a lack of space, but also because the focus here is less on the sum of raw materials and resources that contribute to electronics. A discussion on mining in relation to electronic waste can be found in Grossman, *High Tech Trash.* Mention is also made of this issue in relation to coltan and mobile phones in Nick Couldry and Anna McCarthy, eds., *MediaSpace: Place, Scale, and Culture in a Media Age* (London: Routledge, 2004), 2–3.

60. In my use of the term *circulation,* I am influenced by Dilip Parameshwar Gaonkar and Elizabeth A. Povinelli, who write, "In a given culture of circulation, it is more important to track the proliferating copresence of varied textual/cultural forms in all their mobility and mutability than to attempt a delineation of their fragile autonomy and specificity. Or, it is more important if the purpose, as Michel Foucault long ago suggested, is to move between the seductive sparkle of the 'thing' and the quiet work of the generative matrix—the diagram, as Foucault's acolyte Gilles Deleuze would name this node in the production of life that provides us with the outline of the thing and its excess. This *ethnography of forms,* for want of a better term, can be carried out only within a set of circulatory fields populated by myriad forms" ("Technologies of Public Forms: Circulation, Transfiguration, Recognition," *Public Culture* 15, no. 3 [2003]: 391).

61. United Nations Environment Programme, "Basel Conference Addresses Electronic Wastes Challenge," November 27, 2006, http://www.unep.org/Documents.Multilingual/Default.asp?DocumentID=485&ArticleID=5431&l=en.

62. Greenpeace, "Greenpeace Pulls Plug on Dirty Electronics Companies," May 23, 2005, http://www.greenpeace.org/international/en/press/releases/greenpeace-pulls-plug-on-dirty.

63. Rachel Shabi, "The E-Waste Land," *Guardian,* November 30, 2002.

64. Will Straw, "Exhausted Commodities: The Material Culture of Music," *Canadian Journal of Communication* 25, no. 1 (2000), http://www.cjc-online.ca/index.php/journal/article/viewArticle/1148/1067.

65. William Rathje and Cullen Murphy, *Rubbish! The Archaeology of Garbage* (New York: HarperCollins, 1992).

66. Michael Thompson, *Rubbish Theory: The Creation and Destruction of Value* (Oxford: Oxford University Press, 1979).

67. Walter Moser, "The Acculturation of Waste," in *Waste-Site Stories: The Recycling of Memory,* ed. Brian Neville and Johanne Villeneuve (Albany: State University of New York Press, 2002), 102. Several theorists discuss these generative and dynamic qualities of waste, which will be taken up throughout this study. See also Gay Hawkins, *The Ethics of Waste: How We Relate to Rubbish* (Lanham, MD: Rowman and Littlefield, 2005); John Scanlan, *On Garbage* (London: Reaktion, 2005).

68. Victor Buchli writes (as influenced by Butler), "What is more important probably is not to study the materializations themselves but rather what was wasted towards these rapid and increasingly ephemeral materializations" (introduction to *The Material Culture Reader,* ed. Victor Buchli [Oxford: Berg, 2002], 17).

69. Serres, *Parasite,* 13.

Chapter 1

Donna Haraway, "Cyborgs, Coyotes, and Dogs: A Kinship of Feminist Figurations" and "There Are Always More Things Going on Than You Thought! Methodologies as Thinking Technologies: An Interview with Donna Haraway," conducted in two parts by Nina Lykke, Randi Markussen, and Finn Olesen, in *Haraway Reader,* 338.

1. Jim Fisher, "Poison Valley: Is Workers' Health the Price We Pay for High-Tech Progress?" *Salon,* July 30, 2001, http://archive.salon.com/tech/feature/2001/07/30/almaden1/. As Fisher writes, these "aspiring cities are founded on the reduction of a new precious metal—the computer chip—which in the end is just a metalized piece of sand, or silicon."

2. Albert Borgmann, *Holding On to Reality: The Nature of Information at the Turn of the Millennium* (Chicago: University of Chicago Press, 1999), 144. This basic theory of information, which can be traced to Claude Shannon and Warren Weaver, demonstrates how the "bit," as a switching model, is closely tied to the actual operation of electrical currents. In this sense, material systems and informational systems are interdependent. See Claude E. Shannon and Warren Weaver, *The Mathematical Theory of Communication* (1949; repr., Urbana: University of Illinois Press, 1962); Hayles, *How We Became Posthuman.*

3. Christophe Lécuyer and David C. Brock undertake a material investigation into electronic innovation and suggest that not only are microelectronics "under materialized," but that by focusing on materials entirely new questions about the material networks and ecologies of electronics emerge. See "The Materiality of Microelectronics," *History and Technology* 22, no. 3 (September 2006): 301–25.

4. For a detailed analysis of microchip resources and inputs, see Ruediger Kuehr, German T. Velasquez, and Eric Williams, "Computers and the Environment: An Introduction to Understanding and Managing Their Impacts," in Kuehr and Williams, *Computers and the Environment,* 1–16.

5. For reports on the historic development and dot-com to dot-bomb culture of technology in Silicon Valley, see Christophe Lecuyer, *Making Silicon Valley: Innovation and the Growth of High Tech, 1930–1970* (Cambridge, MA: MIT Press, 2005); Christine A. Finn, *Artifacts: An Archaeologist's Year in Silicon Valley* (Cambridge, MA: MIT Press, 2001).

6. Fisher, "Poison Valley." See also the listings for Santa Clara County in the EPA's National Priorities List, http://yosemite.epa.gov/r9/sfund/r9sfdocw.nsf/WSOState!OpenView&Start=1&Count=1000&Expand=2.29#2.29.

Information on Superfund sites in Silicon Valley in the form of legal documents and records of decision was initially made available through the EPA office in San Francisco. Much of this information is now available at http://www.epa.gov/region09/superfund/superfundsites.html. The Silicon Valley Toxics Coalition also features a "Silicon Valley Toxic Tour" on its Web site, where an interactive map allows users to view the different Silicon Valley Superfund sites. See http://www.svtc.org/site/PageServer?pagename=svtc_silicon_valley_toxic_tour.

7. Cleanup is also not an absolute process, as chemicals potentially break down and migrate in unexpected ways to become newly toxic. See Matt Ritchel, "E.P.A. Takes Second Look at Many Superfund Sites," *New York Times,* January 31, 2003.

8. Intel Corporation, "From Sand to Circuits," 2005, ftp://download.intel.com/museum/sand_to_circuits.pdf.

9. Ibid.

10. Kuehr, Velasquez, and Williams, "Computers and the Environment," 7. As these authors write, "Computer production is materials-intensive; the total fossil fuels used to manufacture one computer, for example, amount to nine times the weight of the actual computer." As David Naguib Pellow and Lisa Sun-Hee Park also outline, "In 1999, on average, the production of an eight-inch silicon wafer required the following resources: 4,267 cubic feet of bulk gases, 3,787 gallons of waste water, 27 pounds of chemicals, 29 cubic feet of hazardous gases, 9 pounds of hazardous waste, and 3,023 gallons of deionized water" (*The Silicon Valley of Dreams: Environmental Injustice, Immigrant Workers, and the High-Tech Global Economy* [New York: New York University Press, 2002], 76–77).

11. Mazurek, *Making Microchips*, ix.

12. Not only is it ironic that a considerable amount of wasted material is required to manufacture dust-free microchips, but electronics also generate dust throughout their life cycles. More will be said about dust as waste in chapter 5, but the dust emitted from computers while in use has also been shown to contain brominated flame retardants (BFRs), substances that have a range of possible deleterious effects for people and environments. Elizabeth Grossman writes at length about BFRs in *High Tech Trash;* see also Alexandra McPherson, Beverley Thorpe, and Ann Blake, "Brominated Flame Retardants in Dust on Computers: The Case for Safer Chemicals and Better Computer Design," June 2004, http://www.electronicstakeback.com.

13. A typical Intel fab is also equipped with automated pods on monorails, which monitor the chip "recipes" and transport heavy loads of wafer fabs. For a demonstration of this process, see the Intel Corporation's video "Invention, Innovation, Investment" at http://intelpr.feedroom.com/.

14. Many of the fab workers in the United States are immigrants and economically disadvantaged women of color. With offshoring of microchip fabrication to Southeast Asia and elsewhere, the profile of fab workers is similar. See Terry Harpold and Kavita Philip, "Of Bugs and Rats: Cyber-Cleanliness, Cyber-Squalor, and the Fantasy-Spaces of Informational Globalization," *Postmodern Culture* 11, no. 1 (2000), http://muse.jhu.edu/journals/pmc/v011/11.1harpold.html. See also Pellow and Park, *Silicon Valley of Dreams.* There is very little comprehensive information about fab workers and rates of illness, but one early study is Damien M. McElvenny et al., *Cancer among Current and Former Workers at National Semiconductor (UK) Ltd., Greenock* (Norwich, UK: Health and Safety Executive Books, 2001).

15. Pellow and Park, *Silicon Valley of Dreams*, 76.

16. Ibid., 86–87.

17. Kenneth Geiser, "The Chips Are Falling: Health Hazards in the Microelectronics Industry," *Science for the People* 17, no. 8 (1985), as cited in Pellow and Park, *Silicon Valley of Dreams*, 76–77. As Werner Rugemer further elaborates, "Building smaller and faster circuits . . . requires the use of more solvents and other chemicals to achieve the necessary requirements for 'clean' components. As the geometries of production decrease, more solvents are needed to wash away ever smaller 'killer particles' that could jam a circuit. Smaller and faster may also mean using even more chemicals" ("The Social, Human, and Structural Costs of High Technology: The Case of Silicon Valley," *Nature, Society, and Thought* 1 [1987]: 149–60).

18. In this way, Esther Leslie asks whether it is "possible to tell history from the standpoint of matter—coal, diamonds, gold, metals, glass, dyes, cellophane" (*Synthetic Worlds: Nature, Art, and the Chemical Industry* [London: Reaktion, 2005], 24).

19. Ibid.

20. See the Semiconductor Industry Association's Web site at http://www.sia-online.org/cs/industry_resources/industry_fact_sheet. The SIA estimates that "In 2005, the semiconductor industry made over 90 million transistors for every man, woman and child on Earth, and by 2010, this number should be 1 billion transistors." See also Grossman, *High Tech Trash,* 4; Pellow and Park, *Silicon Valley of Dreams,* 86.

21. Daniel Bell is one of the most well-known historic commentators on the perceived shift to postindustrial economies. See Daniel Bell, *The Coming of the Post-Industrial Society: A Venture in Social Forecasting* (New York: Basic Books, 1976).

22. Pellow and Park, *Silicon Valley of Dreams,* 86.

23. I will not discuss offshoring at length here; for more information on this topic, see Smith, Sonnenfeld, and Pellow, *Challenging the Chip.*

24. Jack Kilby at Texas Instruments and Robert Noyce at Fairchild Semiconductor filed separate patents for versions of the integrated circuit in 1959 (having each developed versions in 1958). They have both since been recognized as founding "inventors" of the integrated circuit.

25. The terms *microprocessor* and *integrated circuit* are both referred to as chips or "microchips" in this study; in technical terms, *microprocessor* refers to a more sophisticated assemblage of transistors, the "computer on a chip."

26. Leslie Berlin, *The Man Behind the Microchip: Robert Noyce and the Invention of Silicon Valley* (Oxford: Oxford University Press, 2005), 137.

27. Ibid., 138.

28. Gordon Moore cited in ibid., 138.

29. Gordon E. Moore, "Cramming More Components onto Integrated Circuits," *Electronics* 38, no. 8 (April 19, 1965): 114–17. While Moore's Law may have been an inviolable principle up to this point, a number of inquiries make continual speculations about the point at which maximum growth will be reached and when an entirely new structure of growth and development will be required (such as nanotechnology). Gordon Moore offers his own speculation on this point in the essay "No Exponential Is Forever . . . but We Can Delay Forever," *Solid State Circuits Conference Proceedings* 1 (2003): 20–23.

30. Drawing a correlation between technological advance and waste, where ongoing growth in processing power renders old models obsolete, Gordon Moore further notes, "If the auto industry advanced as rapidly as the semiconductor industry, a Rolls Royce would get half a million miles per gallon and it would be cheaper to throw it away than to park it" (cited in "4004: Intel's First Microprocessor," available at http://intelpr.feedroom.com/).

31. Intel produces around 75 percent of microchips worldwide, with the nearest competitor, Advanced Micro Devices, accounting for 23 percent of microchip production. Intel's dominance has recently been challenged by a run of antitrust cases. See "Intel's Antitrust Ruling: A Billion-Euro Question," *Economist,* May 14, 2009.

32. Timothy Mitchell, "Rethinking Economy," *Geoforum* 39, no. 3 (2008): 1116–21. See also Timothy Mitchell, "The Character of Calculability," in *Rule of Experts: Egypt, Techno-Politics, Modernity* (Berkeley: University of California Press, 2002), 80–122.

33. Patrick Haggerty, "Integrated Electronics: A Perspective," in *Management Philosophies and Practices of Texas Instruments* (Dallas: Texas Instruments, 1965), as reprinted in Frederick Seitz and Norman G. Einspruch, *Electronic Genie: The Tangled History of Silicon* (Urbana: University of Illinois Press, 1998), 221.

34. Ibid., 252.

35. Claude E. Shannon, "The Mathematical Theory of Communication," in Shannon and Weaver, *Mathematical Theory of Communication*, 31–35.

36. John Durham Peters suggests just how seductive a universal approach to information became, as meaning was detached and reattached in altered form: "'Communication theory' was explicitly a theory of 'signals' and not of 'significance.' But as the terms diffused through intellectual life—and they did so at violent speed—these provisos were little heeded. 'Information' became a substantive and communication theory became an account of meaning as well as of channel capacity. Indeed, the theory may have seemed so exciting because it made something already quite familiar in war, bureaucracy, and everyday life into a concept of science and technology. Information was no longer raw data, military logistics, or phone numbers; it was the principle of the universe's intelligibility" (*Speaking into the Air: A History of the Idea of Communication* [Chicago: University of Chicago Press, 1999], 23). Hayles points out that with developments in information theory, emphasis was placed on the manipulation of informational patterns at the expense of embodiment. She notes, "Aiding this process was a definition of information, formalized by Claude Shannon and Norbert Wiener, that conceptualized information as an entity distinct from the substrates carrying it. From this formulation, it was a small step to think of information as a kind of bodiless fluid that could flow between different substrates without loss of meaning or form" (*How We Became Posthuman*, xi). In the same study, Hayles writes about the dematerializing drive toward total informatization and about how, "in the face of such a powerful dream, it can be a shock to remember that for information to exist, it must *always* be instantiated in a medium" (13).

37. Marshall McLuhan, *Understanding Media* (Cambridge, MA: MIT Press, 1994), 139.

38. James R. Beniger, *The Control Revolution: Technological and Economic Origins of the Information Society* (Cambridge: Harvard University Press, 1989), 25–26. In this study of information technologies, Beniger addresses the possibility for complete assimilation through digitalization, which he suggests "begins to blur earlier distinctions between the communication of information and its processing . . . as well as between people and machines." "Also blurred," he writes, "are the distinctions among information types: numbers, words, pictures, and sounds, and eventually tastes, odors, and possibly even sensations, all might one day be stored, processed, and communicated in the same digital form. In this way digitalization promises to transform currently diverse forms of information into a generalized medium for processing and exchange by the social system, much as, centuries ago, the institution of common currencies and exchange rates began to transform local markets into a single world economy."

39. Ibid., 23. By locating the "crisis" of production and control in the nineteenth century, Beniger suggests that information overload has a longer history than recent postwar computing devices. Printing, punch cards, and preelectric forms of automation all contributed to a form of overload. Computing, then, emerges as yet another technology that attempted to control the deluge.

40. Ibid., 17–18.

41. Punch cards are the classic example of the ways in which attempts to manage information (e.g., Census data) contributed to the development of new technologies (punch cards) to manage that information. See Jennifer Gabrys, "Paper Mountains,

Disposable Cities," in *Surface Tension Supplement* 1, ed. Brandon Labelle and Ken Ehrlich (Los Angeles: Errant Bodies Press, 2006), 130–39. For another example of these integrated economies of excess, see Abigail Sellen and Richard Harper, *The Myth of the Paperless Office* (Cambridge, MA: MIT Press, 2002).

42. Engineer and information scientist John Robinson Pierce notes that "an increase in possibilities increases entropy" ("The Origins of Information Theory," in *An Introduction to Information Theory: Symbols, Signals, and Noise* [New York: Dover, 1980], 23). Warren Weaver similarly considers how the tendency toward overproduction of new information leads to entropy, or to what information theorists refer to as "noise." Weaver discusses how a channel may be saturated with information to overload the delivery process completely. In this way, "if one tries to overcrowd the capacity of the audience, it is probably true, by direct analogy, that you do not, so to speak, fill the audience up and then waste only the remainder by spilling. More likely, and again by direct analogy, if you overcrowd the capacity of the audience you force a general and inescapable error and confusion." In fact, Weaver correlates the noise, or entropy, of a signal with both the "character of the source" and the "capabilities of the channel" through which it passes. His extended explanation makes clear, however, that the entropy is actually related to the material restraints the medium sets on the message. See Warren Weaver, "Some Recent Contributions to the Mathematical Theory of Communication," in Shannon and Weaver, *Mathematical Theory of Communication*, 1–28.

43. Beniger, *Control Revolution*, 47–48.

44. Jean-François Lyotard, *The Postmodern Condition: A Report on Knowledge*, trans. Brian Massumi (Minneapolis: University of Minnesota Press, 1985), 4. This report was originally commissioned by the Conseil des Universités of the government of Quebec. Arguably, widespread use of electronic information technologies and postmodernism emerge at the same moment for a reason: the perceived excess (of information and speed) that generates from and through these devices gives rise to discourses of proliferation.

45. See Todd Gitlin, *Media Unlimited: How the Torrent of Images and Sounds Overwhelms Our Lives* (New York: Henry Holt, 2003).

46. Peter Lyman and Hal R. Varian, "How Much Information," 2003, http://www2.sims.berkeley.edu/research/projects/how-much-info-2003/index.htm. Information about the ongoing 2008 "How Much Information" study can be found on the Web site of the Global Information Industry Center at the University of California, San Diego, http://hmi.ucsd.edu/howmuchinfo.php.

47. Lyman and Varian, "How Much Information."

48. Ibid.

49. Another private study, sponsored by EMC Corporation, suggests that these numbers are much higher, with as many as 3,892,179,868,480,350,000,000 bits added to the "Digital Universe" in 2008, which would be equivalent to an increase of 487 exabytes (almost 30 times more bytes than estimated by the "How Much Information" study in 2003). See IDC, *The Diverse and Exploding Digital Universe: An Updated Forecast of Worldwide Information Growth through 2011* (Framingham, MA: IDC, 2008).

50. Many computing texts—historic and contemporary—emphasize the basic calculating function of computers. For example, see Gordon Pask and Susan Curran, *Micro Man: Living and Growing with Computers* (London: Century, 1982). Media theorists have also discussed at length the relation between calculation and com-

putation. For example, Darin Barney elaborates on the notion of the calculative episteme of computing technologies in *Prometheus Wired* (Vancouver: University of British Columbia, 2000), 61.

51. Georg Simmel discusses the ways in which quantities can give rise to new qualities in *The Philosophy of Money,* trans. Tom Bottomore and David Frisby (1907; repr; London: Routledge, 1990), 278–80. In this respect, as Don Slater and Andrew Barry emphasize through the ongoing developments in metrology, calculation does not just count or reflect "reality"; instead, it creates "calculable objects" that can give rise to "new realities." See Slater and Barry's introduction to *The Technological Economy,* ed. Don Slater and Andrew Barry (London: Routledge, 2005), 11; Bruno Latour, *Pandora's Hope: Essays on the Reality of Science Studies* (Cambridge: Harvard University Press, 1999).

52. Lyman and Varian, "How Much Information." To give another sense of just how big five exabytes is, the authors elaborate, "If digitized, the nineteen million books and other print collections in the Library of Congress would contain about ten terabytes of information; five exabytes of information is equivalent in size to the information contained in half a million new libraries the size of the Library of Congress print collections."

53. Vannevar Bush, "As We May Think," *Atlantic Monthly,* July 1945, 101–8, available at http://dx.doi.org/10.3998/3336451.0001.101 Similar economies of scale and compression are still at work today, most notably with the ongoing Google Books project. See http://books.google.com/googlebooks/history.html.

54. Bush, "As We May Think."

55. McLuhan, *Understanding Media,* 111.

56. Dave Patterson, "A Conversation with Jim Gray," *ACM Queue* 1, no. 4 (June 2003): 8–17. http://queue.acm.org/detail.cfm?id=864078.

57. I will discuss these qualities of calculation further in chapter 2. See also Michel Callon and John Law, "On Qualculation, Agency and Otherness," *Environment and Planning D: Society and Space* 23, no. 5 (2005): 717–33.

58. As discussed throughout this study, I am here extending N. Katherine Hayles's discussion of the imagining of immaterial information to consider how dematerialization operates beyond the "medium" and extends to infrastructures or media environments. See Hayles, *How We Became Posthuman.*

59. Simmel described this sensation in 1903, at a time when urban stimuli challenged all strategies of adaptation and response. The intensity, speed, and rapid change of impressions found in the modern metropolis, Simmel argued, gave rise to new strategies of sensorial navigation. See Georg Simmel, "The Metropolis and Mental Life," in *Simmel on Culture,* ed. David Frisby and Mike Featherstone (1903; repr., London: Sage, 1997), 174–86.

60. A number of writers have worked through the notion of virtual geography. For example, see William J. Mitchell, *City of Bits: Space, Place, and the Infobahn* (Cambridge, MA: MIT Press, 1995).

61. For a description of this Superfund redevelopment, see Environmental Protection Agency, "Fairchild Semiconductor Case Study," http://epa.gov/superfund/programs/recycle/live/casestudy_fairchild.html.

62. While one finds the most Superfund sites in the United States within the amorphous boundaries of Silicon Valley, there is another register of abundance in this same location: Silicon Valley is also the site of one of the largest concentrations

of millionaires in the world (see Pellow and Park, *Silicon Valley of Dreams,* 1). Many of these millionaires, however, refer to themselves as "working-class millionaires," given the relative poverty they feel they experience in comparison to highly paid executives. See Gary Rivlin, "In Silicon Valley, Millionaires Who Don't Feel Rich," *New York Times,* August 5, 2007.

63. In a Marxian analysis, David Harvey discusses the role of spaces of accumulation and suggests that the growth of economies requires "the necessary creation of a geographical landscape to facilitate accumulation through production and circulation" (*Spaces of Capital: Towards a Critical Geography* [London: Routledge, 2001], 266).

64. Ibid., 237.

65. The need to renew the tools of production constantly, as Marx has stated, contributes to the inevitable production of waste. This is clearly demonstrated in Silicon Valley (and beyond) by the fact that Intel amortizes over 1 billion dollars in infrastructural costs annually in order to remain at the cutting edge of microchip production. See Karl Marx, *Capital: A Critique of Political Economy,* vol. 1, trans. Ben Fowkes (1976; repr., London: Penguin, 1990); Gordon Moore, interview, March 3, 1995, *Silicon Genesis: An Oral History of Semiconductor Technology,* Stanford and the Silicon Valley Project, http://silicongenesis.stanford.edu/complete_listing.html. Moore's Law, as it turns out, may level off not due to physical constraints but because of the economic constraints imposed by the costs of fabs. Smaller circuitry will become increasingly expensive to manufacture, with some technologists estimating that by 2014, when circuitry dimensions are expected to be as small as 18 nanometers, fab costs will outstrip profits recovered. See Jack Schofield, "When the Chips Are Down," *Guardian,* July 29, 2009.

66. Pellow and Park, *Silicon Valley of Dreams,* 16–18.

67. Jean Baudrillard, "The Remainder," in *Simulacra and Simulation,* trans. Sheila Faria Glaser (Ann Arbor: University of Michigan Press, 1994), 145.

Chapter 2

Donna Haraway, "A Cyborg Manifesto: Science, Technology, and Socialist-Feminism in the Late Twentieth Century," in *Simians, Cyborgs, and Women: The Reinvention of Nature* (New York: Routledge, 1991), 153.

Serres, *Parasite,* 52.

1. For documentation of this project, see Dennis Oppenheim, interview, March 29, 1969, in *Recording Conceptual Art,* ed. Alexander Alberro and Patricia Norvell (Berkeley: University of California Press, 2001), 21–30.

2. According to the NASDAQ Newsroom "Performance Report," "NASDAQ is the largest U.S. electronic stock market. With approximately 3,200 companies, it lists more companies and, on average, trades more shares per day than any other U.S. market. It is home to companies that are leaders across all areas of business, including technology, retail, communications, financial services, transportation, media and biotechnology" (http://www.nasdaq.com/newsroom/stats/Performance_Reports.stm, accessed March 4, 2008).

3. NASDAQ, "Built for Business: Annual Report," 2004. The "2007 Annual Report" indicates that "in 2007, NASDAQ became the largest U.S. equities exchange by volume with an average in NASDAQ securities of 2.17 billion shares per day"

trading on the NASDAQ platform. See NASDAQ, "2007 Annual Report," 7, http://ir.nasdaq.com/annuals.cfm.

4. The term *new economy* is much debated and discussed. For more information on the term and phenomenon, see Doug Henwood, *After the New Economy: The Binge and the Hangover That Won't Go Away* (New York: New Press, 2005); Jean Gadrey, *New Economy, New Myth* (2001; London: Routledge, 2003); Manuel Castells, *The Rise of the Network Society* (Oxford: Blackwell, 2000). Lisa Adkins discusses the particular ways in which the "new economy" can be seen to give rise to altered relations between people and property and to definitions of personhood and labor, in "The New Economy, Property, and Personhood," *Theory, Culture & Society* 22, no. 2 (2005): 111–30.

5. Melissa Fisher and Greg Downey elaborate on this point: "The New Economy looked like an old story: speculator hype driving a stock market bubble. . . . There was no New Economy, if that term meant the dawn of an age without business contractions, where the cyclical laws of the economy had been repeated" ("Introduction: The Anthropology of Capital and the Frontiers of Ethnography," in *Frontiers of Capital: Ethnographic Reflections on the New Economy*, ed. Melissa Fisher and Greg Downey [Durham: Duke University Press, 2006], 2).

6. Thompson argues this point throughout *Rubbish Theory*, although he discusses the migration of value across categories of value, devaluation, and revaluation primarily in relation to houses and antiques.

7. To this extent, as Harvey writes when elaborating on the relation between crisis and saturated markets so often discussed by Marx, "Periodic crises must in general have the effect of expanding the productive capacity and renewing the conditions of further accumulation. We can conceive of each crisis as shifting the accumulation process onto a new and higher plane" (Harvey, *Spaces of Capital*, 241).

8. For a more extensive discussion on the stock ticker as a "generator" of temporalities, see Alex Preda, "Socio-Technical Agency in Financial Markets: The Case of the Stock Ticker," *Social Studies of Science* 36, no. 5 (October 2006): 753–82.

9. My use of the performative here draws on literature that discusses the more affective and material aspects of markets, as well as analyses of the ways in which markets perform and so constitute economic conditions. See, for instance, Nigel Thrift, "Performing Cultures in the New Economy," *Annals of the Association of American Geographers* 90, no. 4 (2000): 674–92; Michael Pryke and John Allen, "Monetized Time-Space: Derivatives—Money's 'New Imaginary?'" *Economy and Society* 29, no. 2 (May 2000): 264–84; Donald MacKenzie, Fabian Muniesa, and Lucia Siu, eds., *Do Economists Make Markets? On the Performativity of Economics* (Princeton: Princeton University Press, 2007).

10. Donald MacKenzie, "Is Economics Performative? Option Theory and the Construction of Derivative Markets," and Michel Callon, "What Does It Mean to Say That Economics Is Performative?" in MacKenzie, Muniesa, and Siu, *Do Economists Make Markets?* 54–86, 311–57.

11. Benjamin sought to describe this more phenomenal aspect of "economy and culture" through his readings (in some ways, against the grain) of Marx. As Benjamin suggests, he was less interested in the "causal" or originary aspects of economies and more attentive to their "expressive" aspects. This expressive element could even influence the shape of theories that describe those economies. So Benjamin asserts, in his Convolute on the stock exchange, that his study of the arcades "will demonstrate how the milieu in which Marx's doctrine arose affected that doc-

trine through its expressive character (which is to say, not only through causal connections); but, second, it will also show in what respects Marxism, too, shares the expressive character of the material products contemporary with it." See Benjamin, *Arcades Project*, 460. The present study on electronic waste similarly seeks to draw out such shared elements of economy and culture, particularly (but not exclusively) through readings of postmodern and media theory as corresponding in expression with information economies.

12. Rita Raley, "eEmpires," *Cultural Critique* 57 (Spring 2004): 121–22. In fact, electronic communication networks now constitute a considerable portion of online trading, and these private mechanisms often enable 24-hour trading. But the full scope of electronic markets and technologies cannot even be circumscribed with these networks. As discussed by Pryke and Allen in "Monetized Time-Space," financial deregulation has, of course, played a significant role in the alteration of markets.

13. Raley, "eEmpires," 111.

14. To this end, media theorist Nick Dyer-Witheford notes that the growth of digital technology is "inseparable" from the growth of these markets. See Dyer-Witheford, *Cyber-Marx: Cycles and Circuits of Struggle in High-Technology Capitalism* (Urbana: University of Illinois Press, 1999), 139.

15. See Bette K. Fishbein et al., *Extended Producer Responsibility: A Materials Policy for the 21st Century* (New York: Inform, 2000); Bette K. Fishbein, *Waste in a Wireless World: The Challenge of Cell Phones* (New York: Inform, 2002). When assessing trends toward dematerialization, some researchers have argued that material flows should be evaluated not just in terms of mass or weight but also in terms of the impact—or "environmental weights"—of materials. Some materials may have a relatively small bulk, but their environmental weight may be considerably larger. This is particularly true for metals. See Ester van der Voet, Lauran van Oers, and Igor Nikolic, "Dematerialization: Not Just a Matter of Weight," *Journal of Industrial Ecology* 8, no. 4 (2005): 121–37.

16. N. Currimbhoy, "New NASDAQ MarketSite Design Inspired by Computer Chip (New-York-City Headquarters Resembles the Inside of a Computer," *Architectural Record* 186, no. 5 (May 1998): 262. C & J Partners of Pasadena, California, was the primary design firm responsible for MarketSite, while Illuminating Concepts completed the lighting design.

17. Ibid.

18. NASDAQ, "MarketSite Fact Sheet," http://www.nasdaq.com/reference/marketsite_facts.stm.

19. Anthony Dunne, *Hertzian Tales: Electronic Products, Aesthetic Experience, and Critical Design* (Cambridge, MA: MIT Press, 2005), 26–27.

20. For extended studies on the virtual, see N. Katherine Hayles, *My Mother Was a Computer* (Chicago: University of Chicago Press, 2005); Brian Massumi, *Parables for the Virtual: Movement, Affect, Sensation* (Durham: Duke University Press, 2002); Elizabeth Grosz, *Architecture from the Outside* (Cambridge, MA: MIT Press, 2001).

21. See James G. Carrier and Daniel Miller, *Virtualism: A New Political Economy* (Oxford: Berg, 1998).

22. Indeed, Bernard Stiegler suggests that "the so-called 'financial bubble' is becoming autonomous to such an extent that it is often cut off from productive realities, and functions according to a logic of belief (or of credit) *massively determined by the performance of telecommunication and computer-based systems* in the management

of financial data. Capital exchanges have become a problem of informational management effected 'in a nanosecond.' These exchanges are data that are exchanged and processed, and no longer monetary masses. Decisions made 'in a nanosecond' are calculations performed on series of indicators dealing primarily with the stock markets themselves and with macroeconomic decisions interfering with them, and not evaluations of the macroeconomic situations of particular enterprises." In this sense, Stiegler writes, "This is the context in which the limited company and the stock-market system developed, with the aim of assuring the mobility of capital" (*Technics and Time*, vol. 1, *The Fault of Epimetheus*, trans. Richard Beardsworth and George Collins [Stanford: Stanford University Press, 1998], 38). Rather than argue that bubbles are "cut off" from reality, however, the present study suggests that the performance of electronic technologies contributes to and mobilizes reality effects, in both the rise and fall of values.

23. Intensity of market practices is one register through which market performativity is now increasingly described. See Scott Lash and Celia Lury, *Global Culture Industry* (Cambridge: Polity, 2007); Karin Knorr-Cetina and Urs Bruegger, "Global Microstrucures: The Virtual Societies of Financial Markets," *American Journal of Sociology* 107, no. 4 (2002): 905–50.

24. Nigel Thrift, *Knowing Capitalism* (London: Sage, 2005), 126.

25. Thrift goes so far as to state that such expenditure, which came to define the new economy, in fact constituted "a kind of forced technological march" (ibid., 127).

26. Robert J. Shiller, *Irrational Exuberance* (Princeton: Princeton University Press, 2000), 21. Elsewhere, Michael Callon suggests, "To predict economic agents' behaviors an economic theory does not have to be true; it simply needs to be believed by everyone" ("What Does It Mean to Say That Economics Is Performative?" 322).

27. Shiller, *Irrational Exuberance*, 29. Caitlin Zaloom also discusses this overlay of financial news and trading in *Out of the Pits: Traders and Technology from Chicago to London* (Chicago: Chicago University Press, 2006).

28. NASDAQ, "MarketSite Fact Sheet."

29. Gordon L. Clark and Nigel Thrift, "Performing Finance: The Industry, the Media, and Its Image," *Review of International Political Economy* 11, no. 2 (May 2004): 290.

30. Thrift, *Knowing Capitalism*, 124. See also Benjamin Lee and Edward LiPuma. "Cultures of Circulation: The Imaginations of Modernity," *Public Culture* 14, no. 1 (2002): 191–213.

31. Clark and Thrift, "Performing Finance," 298.

32. NASDAQ, "2007 Annual Report," 7.

33. Ibid.

34. Shiller, *Irrational Exuberance*, 39–40. As Shiller elaborates in his discussion of "irrational exuberance," the rate of turnover for NASDAQ stocks rose "from 88% in 1990 to 221% in 1999."

35. Pryke and Allen, "Monetized Time-Space," 270. Benjamin has also made this observation in relation to nineteenth-century economies (drawing on Marx's discussion of the spinning jenny): "More than a hundred years before it was fully manifest, the colossal acceleration of the tempo of living was heralded in the tempo of production. And, indeed, in the form of the machine. . . . The tempo of machine operation effects changes in the economic tempo" (*Arcades Project*, 394).

36. Caitlin Zaloom, "Ambiguous Numbers: Technology and Trading in Global Financial Markets," *American Ethnologist* 30, no. 2 (2003): 259.

37. Ibid.

38. Ibid., 261. Knorr-Cetina and Bruegger emphasize the ways in which screens even qualify the dominant model of markets as networks. They argue instead for a "scopic" understanding of screen-based markets, because "networks are sparse social structures, and it is difficult to see how they can incorporate the patterns of intense and dynamic conversational interaction, the knowledge flows, and the temporal structuration" of market activity. As this study on electronic waste suggests, however, networks are more than organizing conceptual structures and can be understood as part of the processes of electronic materialization, with which screens are continuous and constitutive. See Knorr-Cetina and Bruegger, "Global Microstructures," 910.

39. Donna Haraway, "Cyborgs to Companion Species: Reconfiguring Kinship in Technoscience," in *Haraway Reader,* 303.

40. In this way, media theorist Ludwig Pfeiffer undertakes a study into the elusive "materiality of communication" and maintains that "the fall of matter and materialism does not lead to the immaterial pure and simple; rather, it *branches into* the immaterial *and* its material 'sites' or 'supports.' Instead of substantial objects and their meanings, we get information overload and a new hardness of 'supporting' materials, and new 'performativity' of things and bodies" ("The Materiality of Communication," in *Materialities of Communication,* ed. Hans Ulrich Gumbrecht and K. Ludwig Pfeiffer, trans. William Whobrey [Stanford: Stanford University Press, 1994], 2).

41. Hayles, *How We Became Posthuman,* 28. As Hayles writes, "Information in fact derives its efficacy from the material infrastructures it appears to obscure. This illusion of erasure should be the *subject* of inquiry, not a presupposition that inquiry takes for granted." Hayles contests the supposed separation between materiality and information and further argues, "The point of highlighting such moments is to make clear how much had to be erased to arrive at such abstractions in all theorizing, for no theory can account for the infinite multiplicity of our interactions with the real." Here she reveals how the abstract model works to erase multiplicity and stands in for the real such that noise appears as an intrusion rather than an actual condition that challenges any pretense toward complete abstraction.

42. Don Slater, "Markets, Materiality, and the 'New Economy,'" in *Market Relations and the Competitive Process,* ed. Stan Metcalfe and Alan Warde (Manchester: Manchester University Press, 2002), 95–113. See also Slater and Barry, *Technological Economy.*

43. Slater, "Markets, Materiality, and the 'New Economy.'"

44. For a more detailed history of the first computers used to control stock, inventory, and even pastries, see Mike Hally, *Electronic Brains* (London: Granta, 2005).

45. Kevin Kelly, *Out of Control: The New Biology of Machines, Social Systems, and the Economic World* (New York: Basic Books, 1994), 186.

46. Ibid., 189.

47. Ibid.

48. The massive server farms that power NASDAQ are alone evidence of this considerable resource base, in addition to the other examples cited up to this point. Servers—for everything from NASDAQ to Google to the Internet in general—are a major source of energy consumption (and carbon dioxide emissions). See Bobbie Johnson, "Web Providers Must Limit Internet's Carbon Footprint, Say Experts," *Guardian,* May 3, 2009.

49. Dan Schiller, *Digital Capitalism: Networking the Global Market System* (Cambridge, MA: MIT Press, 1999), 16.

50. Ibid., 15.

51. Ibid., 16–17. As Schiller writes, "Inclusive of computing and telecommunications, information technology was proclaimed (by the American Electronics Association) the United States' largest industry."

52. Ibid., 17. Barney also writes on the effectiveness of networks as engines of economic growth: "Computer networks did not create the globalized, privatized economy; they do, however, make it possible to exploit this economy. Flexible production and mass customization are based on speedy, but controlled, processing of systems information, a requirement that the movement of bits over networked computers is particularly suited to meet. . . . Networks are the essential technology for those 'agile' and 'virtual' enterprises that are 'thriving on change and uncertainty' in the era of perfecting capitalism" (*Prometheus Wired*, 130).

53. Raley writes that this "Electronic Empire" "is a loose assemblage of relations characterized by another set of terms: flexibility, functionality, mobility, programmability, and automation. The paradigm for such an assemblage is the network, which involves new geopolitical orderings, a reconfigured sense of center and periphery and an attendant complication of the world-system idea" ("eEmpires," 132).

54. Kelly, *Out of Control*, 193.

55. Schiller, *Digital Capitalism*, xv. Algorithmic trading, or "exchange automation" is also one key way in which the particular functionalities and speeds of market exchanges are performed. See Fabian Muniesa, "Assemblage of a Market Mechanism," *Journal of the Center for Information Studies* 5, no. 3 (2004): 11–19.

56. Kelly writes, "In network economics the major expense of new product development stems from *designing the manufacturing process* and not designing the product" (*Out of Control*, 196).

57. Nigel Thrift and Shaun French, "The Automatic Production of Space," *Transactions of the Institute of British Geographers* 27, no. 3 (2002): 310.

58. Friedrich Kittler, "There Is No Software," *CTHEORY*, October 18, 1995, http://www.ctheory.net/articles.aspx?id=74.

59. Ibid. Kittler also writes, "Only in Turing's paper *On Computable Numbers with an Application to the Entscheidungsproblem* there existed a machine with unbounded resources in space and time, with infinite supply of raw paper and no constraints on computation speed. All physically feasible machines, in contrast, are limited by these parameters in their very code. The inability of Microsoft DOS to tell more than the first eight letters of a file name such as WordPerfect gives just a trivial or obsolete illustration of a problem that has provoked not only the ever-growing incompatibilities between the different generations of eight-bit, sixteen-bit and thirty-two-bit microprocessors, but also a near impossibility of digitizing the body of real numbers formerly known as nature."

60. John Diebold, "Factories without Men: New Industrial Revolution," *Nation*, September 19, 1953, 227–28, 250–51, 271–72. See also Paul Ceruzzi, *A History of Modern Computing* (Cambridge, MA: MIT Press, 2003), 32.

61. John Diebold, *Automation: The Advent of the Automatic Factory* (New York: D. Van Nostrand, 1952), 46.

62. Ibid., 46–47.

63. Ibid., 47.

64. Ibid., 30. In many ways, this optimization and improved efficiency in the production process simply shifted the problem of waste to appear to be a problem related to consumption. More will be said about this in the following chapters, which address disposal and obsolescence.

65. Ibid., 32.

66. Marshall McLuhan, "Automation: Learning a Living," in *Understanding Media*, 347.

67. Michel Serres, *Hermes: Literature, Science, Philosophy*, ed. Josue V. Harari and David F. Bell (Baltimore: Johns Hopkins University Press, 1982), 56.

68. Hansen, *Embodying Technesis*, 61.

69. Serres further writes, "At least by the change of directions, at least by the division of flows, by bifurcation, at least by semiconduction, one-way streets and no entries, at least by orientation. Hermes is the god of the crossroads, and is the god of whom Maxwell made a demon. Thus the message, passing through his hands in the location of the exchanger, is changer. It arrives neither pure nor unvarying nor stable" (*Parasite*, 42–43).

70. Arjun Appadurai, "Introduction: Commodities and the Politics of Value," in *The Social Life of Things: Commodities in Cultural Perspective*, ed. Arjun Appadurai (Cambridge: Cambridge University Press, 1986), 4. Appadurai suggests that we explore "the conditions under which economic objects circulate in different *regimes of value* in space and time."

71. Ibid., 9.

72. Thrift, *Knowing Capitalism*, 7.

73. For more on such "refashioning" of the commodity, see Haraway, *Modest_Witness@Second_Millenium*, 142.

74. Fritz Machlup, *Knowledge: Its Creation, Distribution, and Economic Significance*, vol. 1, *Knowledge and Knowledge Production* (Princeton: Princeton University Press, 1980), 162. Such "flows of knowledge" took actual form in a 1949 analog computer designed by Bill Phillips, which modeled varying economic conditions through flows of differently colored water. Phillips designed this computer while at the London School of Economics, and the machine is now held by the Science Museum of London.

75. Ibid., 162.

76. Ibid., 171. As Machlup notes, "The production and distribution of knowledge in the United States is, in essence, the annual flow of knowledge disseminated at a cost (defrayed or borne by some members of our society)."

77. Kelly, *Out of Control*, 209–10.

78. Spam has reached such a volume that "pump and dump" strategies for promoting and then selling shares can at times have a significant effect on markets. See "Spammers Manipulate Stock Markets," *BBC News*, August 25, 2006, http://news.bbc.co.uk/go/pr/fr/-/2/hi/technology/5284618.stm.

79. The proportion of e-mail traffic that consists of spam reportedly ranges from 50 to 80 percent of all Internet traffic. Similarly, the proportion of U.S. mail that constitutes "junk" or bulk mailings is as much as 50 percent. See Andrew Odlyzko, "The History of Communications and Its Implications for the Internet," 2000, http://www.dtc.umn.edu/~odlyzko/doc/history.communications0.pdf.

80. Paul Reyes details the remainders of these recently failed economies in "Bleak Houses: Digging through the Ruins of the Mortgage Crisis," *Harper's,* October 2008, 31–45.

81. Michael Shanks, David Platt, and William L. Rathje, "The Perfume of Garbage: Modernity and the Archaeological," *Modernism/Modernity* 11, no. 1 (2004): 72.

82. John Frow, "Invidious Distinction: Waste, Difference, and Classy Stuff," in *Culture and Waste: The Creation and Destruction of Value,* ed. Gay Hawkins and Stephen Muecke (Lanham, MD: Rowman and Littlefield, 2003), 35.

83. Ibid., 36.

84. Design theorist Anne-Marie Willis addresses some of the contradictory notions of dematerialization: "While initially, information technology seemed to offer obvious opportunities for impact reduction (e.g., less need to travel, less need for paper) it in fact has provided endless possibilities for driving *new* forms of material throughput." "De/re/materialization (contra-futures)," *Design Philosophy Papers* 2 (2005), http://www.desphilosophy.com.

85. The World Resources Institute report on material flows documents the relatively constant quantities of resources in use within manufacturing, even with the trend toward fewer resource inputs per unit manufactured. See Allen Hammond et al., *Resource Flows: The Material Basis of Industrial Economies* (Washington, DC: World Resources Institute, 1997).

86. Jean-Pierre Dupuy, "Myths of the Informational Society," in *The Myths of Information: Technology and Postindustrial Culture,* ed. Kathleen Woodward (Madison, WI: Coda, 1980), 6. Dupuy writes, "Whether they talk of 'new growth' or a 'new international economic order,' the underlying strategy is the same: let's send our heavy industries abroad where they can pollute the countries of the Third World, spoiling *their* landscapes, deadening *their* workers, and disrupting *their* time and space, and let's keep for ourselves the growth of immaterial productions which do not poison the atmosphere, are suited to decentralized locations, and enable us to solve to a large extent the problem of unemployment."

87. Lisa Parks, "Kinetic Screens: Epistemologies of Movement at the Interface," in *MediaSpace: Place, Scale and Culture in a Media Age,* ed. Nick Couldry and Anna McCarthy (London: Routledge, 2004), 52.

88. Cathode-ray tubes, commonly found in monitors for computers and televisions, contain lead oxide and cadmium, substances that are toxic to humans (and the environment). When recycled, the copper tubes at the end of the CRT yokes are broken off and sold for metal recovery. CRTs in monitors are now increasingly replaced by liquid crystal displays. There is some speculation that due to new flat-panel screens and the introduction of digital television, television and computer screen disposal may increase considerably. See Basel Action Network and Silicon Valley Toxics Coalition, *Exporting Harm,* 5.

89. Haraway, "Cyborg Manifesto," 153.

Chapter 3

J. David Bolter, *Turing's Man: Western Culture in the Computer Age* (Chapel Hill: University of North Carolina Press, 1984), 121.

1. Italo Calvino, *Invisible Cities,* trans. William Weaver (New York: Harcourt Brace Jovanovich, 1974), 91. Calvino writes, "The bulk of the outflow increases and the

piles rise higher, become stratified, extend over a wider perimeter. Besides, the more Leonia's talent for making new materials excels, the more the rubbish improves in quality, resists time, the elements, fermentations, combustions."

2. Vance Packard, *The Waste Makers* (New York: David McKay, 1960), 4–5.

3. International Association of Electronics Recyclers, *IAER Electronics Recycling Industry Report*, 7.

4. Basel Action Network, *Digital Dump*, 12.

5. Appadurai, "Introduction," 5.

6. Robin Murray, *Zero Waste* (London: Greenpeace Environmental Trust, 2002). Murray cites the 1997 World Resources Institute study already cited in chapter 2: "The World Resources Institute led an international team that traced the flows of 55 materials in 500 use streams (covering 95% of the weight of materials in the economy) for four leading OECD economies (the USA, Japan, the Netherlands and Germany). They found that the total materials requirement in these countries was 45 to 85 metric tonnes per person and that of this between 55% and 75% takes the form of waste materials that are discarded in the course of production (such as mining overburden, agricultural waste or material removed for infrastructural works)." See also Hammond, *Resource Flows*.

7. Rathje and Murphy, *Rubbish!* 188–89.

8. Rudi Colloredo-Mansfeld, "Introduction: Matter Unbound," *Journal of Material Culture* 8, no. 3 (2003): 246.

9. While early studies on electronic waste focused on PCs and mobile phones, the full delineation has now been extended to include this range of electrical and electronic devices. For a discussion on the early estimates of electronic waste volumes, see H. Scott Matthews and Deanna Matthews, "Information Technology Products and the Environment," in Kuehr and Williams, *Computers and the Environment*, 17–40.

10. Intel is "inside" just about every device imaginable. With "Intel Inside," and with chips that continually ramp up every 18 months, not just computers but also every manner of electronic appliance and gadget become subject to this temporal trajectory for innovation. On the far-reaching effects of "Intel Inside," Harvey Molotch writes, "In some cases, the same source provides identical or near-identical elements for many purveyors. 'Intel's Inside' a lot of different products that compete with one another but which get their chip from the same producer—Intel" (*Where Stuff Comes From: How Toasters, Toilets, Cars, Computers, and Many Other Things Come to Be as They Are* [New York: Routledge, 2003], 205–6).

11. Vernon L. Fladager, *The Selling Power of Packaging* (New York: McGraw-Hill, 1956), 94.

12. Katie Dean, "Disposable DVDs at Crossroads," *Wired News*, February 7, 2005, http://www.wired.com/entertainment/music/news/2005/02/66513.

13. Amanda Onion, "Buy, Use, Dispose: A Spike in Disposable Products Has Environmentalists Worried," *ABC News*, December 4, 2002.

14. Packard, *Waste Makers*, 54.

15. This logic is taken up by a number of (postmodern) theorists, including Baudrillard: "What is produced today is not produced for its use-value or its possible durability, but rather with an *eye to its death*, and the increase in the speed with which that death comes about is equaled only by the speed of price rises. . . . Now, we know that the order or production only survives by paying the price of this

extermination, this perpetual calculated 'suicide' of the mass of objects, and that this operation is based on technological 'sabotage' or organized obsolescence under cover of fashion" (*The Consumer Society: Myths and Structures,* trans. Chris Turner [London: Sage, 1998], 46).

16. Gay Hawkins, "Plastic Bags: Living with Rubbish," *International Journal of Cultural Studies* 4, no. 1 (2001): 9. As Benjamin writes, "The dialectic of commodity production in advanced capitalism: the novelty of products—as a stimulus to demand—is accorded an unprecedented importance. At the same time, the 'eternal return of the same' is manifest in mass production" (*Arcades Project,* 331).

17. To this extent, geographer Kevin Hetherington calls for the study of consumption and disposal as related practices. He writes, "Studying consumption makes no sense unless we consider the role of disposing as an integral part of the totality of what consumer activity is all about." See Kevin Hetherington, "Secondhandedness: Consumption, Disposal, and Absent Presence," *Environment and Planning D: Society and Space* 22, no. 1 (2004): 158.

18. Diebold's writings ("The Diebold Group, Inc., Client Reports, 1957–1990," held at the Charles Babbage Institute, University of Minnesota) provide an extensive record of how material goods have been transformed—and have also proliferated—through "computerized" automation. Diebold consulted to companies ranging from General Electric to Time Incorporated and K-mart. In the process, he established a clear logic about how computerized automation would contribute to the growth of economies.

19. Hawkins, "Plastic Bags," 9.

20. With new levels of convergence and pervasive computing with wireless technologies, electronic waste may increase even further. For an extended discussion of this dilemma, see Andreas Köhler and Lorenz Erdmann, "Expected Environmental Impacts of Pervasive Computing," *Human and Ecological Risk Assessment* 10, no. 5 (October 2004): 831–52.

21. Ezio Manzini, *The Material of Invention* (London: Design Council, 1986), 29. Manzini further writes, "The only way to describe the material is to consider it as a system capable of performance: thus we shall speak of a 'material,' not by defining 'what it is,' but describing 'what it does.'"

22. Bernadette Bensaude-Vincent and Isabelle Stengers, *A History of Chemistry,* trans. Deborah van Dam (Cambridge: Harvard University Press, 1996), 205–6.

23. "Molded Plastic Containers," *Modern Packaging Journal* 31 (1957): 120.

24. According to Rathje and Murphy, however, packaging accounts for a comparatively moderate proportion of municipal solid waste in the United States, and they suggest that packaging has actually helped to reduce some types of waste, including food waste. See Rathje and Murphy, *Rubbish!* 216–20.

25. "Molded Plastic Containers," 120.

26. Ellen Lupton and J. Abbott Miller, *The Bathroom, the Kitchen, and the Aesthetics of Waste: A Process of Elimination* (Cambridge: MIT List Visual Arts Center, 1992), 65–66.

27. Adrian Forty, *Objects of Desire: Design and Society, 1750–1980* (London: Thames and Hudson, 1986), 190–93. In this sense, Forty suggests that electricity—particularly as packaged in the form of electrical appliances—depended on "ideas and potential rather than present realities in its appeal to domestic consumers."

28. Roland Barthes, "Plastic," in *Mythologies,* trans. Annette Lavers (New York: Farrar, Straus and Giroux, 1972), 97.

29. Ibid.

30. Ibid.

31. Jeffrey L. Meikle, *American Plastic: A Cultural History* (New Brunswick, NJ: Rutgers University Press, 1995), 299.

32. Ibid.

33. Manzini, *Material of Invention,* 31.

34. Ibid., 32.

35. Ibid., 39.

36. Even though they increasingly appear to be disposable, electronics are not typically designed for disassembly. Electronics recyclers often point out the difficulty of salvaging machines that are not designed with standardized disassembly in mind, which means that most machines must be stripped by hand. Electronic waste then constitutes a formidable waste problem, not least because there is no established or systematic infrastructure to handle this type and volume of waste.

37. The "Electronic Waste Guide" (http://www.ewaste.ch/) suggests that the hazards of electronic waste can be contained if properly handled: "The formation or discharge of hazardous emissions during the recycling of electrical and electronic equipment depends highly on the handling of electronic waste. Hence hazardous substances contained in computers and televisions don't lead automatically to a risk for the environment and the human health."

38. Alvin Toffler, *Future Shock* (New York: Random House, 1970), 55.

39. As John Scanlan writes on Zygmunt Bauman's notion of "fluid modernity," mobile technologies such as these enable a particular kind of "flushing," a sense of fluidity that enables transience and disposability. Such flushing makes way for future movement, and in this respect, "the *liquidity* of the present describes the dematerialization of the object world, not to mention the practical ease with which we can apply the technological flush." Flushing and the fluidity of movement are continuous with the dissipation of materials. See Scanlan, *On Garbage,* 127; Zygmunt Bauman, *Liquid Modernity* (Cambridge: Polity, 2000), 2–14.

40. For a general discussion of dirt and systems, see Mary Douglas, *Purity and Danger: An Analysis of Concepts of Pollution and Taboo* (1966; repr., London: Routledge, 1995).

41. Hetherington, "Secondhandedness," 160.

42. Shanks, Platt, and Rathje, "Perfume of Garbage," 80.

43. Discussing these processes of delay and postponement, cultural historian Susan Strasser writes, following on Mary Douglas, "Sorting and classification have a spatial dimension: this goes here, that goes there. Nontrash belongs in the house; trash goes outside. Marginal categories get stored in marginal places (attics, basements, and outbuildings), eventually to be used, sold, or given away" (Strasser, *Waste and Want* [New York: Henry Holt, Metropolitan, 1999], 6).

44. Hetherington, "Secondhandedness," 162.

45. Ibid., 160.

46. Environmental Protection Agency, "Waste Wise Update: Electronics Reuse and Recycling."

47. Giles Slade documents just how limited existing waste infrastructures would be in their capacity to remove the masses of e-waste in storage. He writes, "But more practically, the e-waste problem will soon reach such gigantic proportions that it will overwhelm our shipping capacity. The world simply cannot produce enough *containers* for America to continue at its current level as an exporter of both

electronic goods and electronic waste" (*Made to Break: Technology and Obsolescence in America* [Cambridge: Harvard University Press, 2006], 3).

48. Hetherington, "Secondhandedness," 169.

49. The process of destroying data on hard drives is notoriously difficult, and many electronics recyclers first set up business to deal not with the reuse of machines but, rather, with the destruction of sensitive data on computers. Typically, recyclers will offer a certification that ensures the elimination of data from hard drives. Many computers, however, are "recycled" without this guarantee, and as a result, hard drives have been scoured for personal data. The Basel Action Network documents leftover sensitive data on hard drives sent from the United States to Africa in their report *The Digital Dump* and in media related to this project on their Web site, http://www.ban.org.

50. This recycling experiment was conducted at a time when relatively few channels existed for the recycling of consumer electronics. These programs have since become better established, although they are not without their difficulties. More will be said about electronics recycling in chapter 5 and the conclusion.

51. Basel Action Network and Silicon Valley Toxics Coalition, *Exporting Harm*, 6.

52. The WEEE Directive is the first of such systematic attempts to prevent the flow of electronics to landfills. Initiated by the European Union, the WEEE Directive has recently required that manufacturers and producers take responsibility (extended producer responsibility, or EPR) for electronics at end of life by providing options for take-back and recycling of their machines. The program has had difficulties with enforcement, however. As a result, some commentators suggest the WEEE and RoHS regulations have become prime examples of "greenwash." See Fred Pearce, "Greenwash: WEEE Directive Is a Dreadful Missed Opportunity to Clean up E-waste," *Guardian*, June 25, 2009.

53. See Fishbein et al., *Extended Producer Responsibility*.

54. Up to 80 percent of electronics shipped to Asia for recycling have their point of origin with raw material brokers in the United States. See Basel Action Network and Silicon Valley Toxics Coalition, *Exporting Harm*, 1–2. See also Jennifer Clapp, *Toxic Exports* (Ithaca: Cornell University Press, 2001), for an in-depth analysis of the geopolitical implications of the movements of hazardous waste.

55. "Greenpeace Deploys GPS to Track Illegal Electronic Waste," Environment Blog, *Guardian*, February 18, 2009, http://www.guardian.co.uk/environment/blog/2009/feb/18/greenpeace-electronic-waste-nigeria-tv-gps; "Undercover Operation Exposes Illegal Dumping of E-waste in Nigeria," *Greenpeace News*, February 18, 2009, http://www.greenpeace.org/international/en/news/features/e-waste-nigeria180209.

56. See Basel Action Network, *Digital Dump;* Fishbein, *Waste in a Wireless World*.

57. Harvey, *Spaces of Capital*, 252–53.

58. Dan Glaister, "US Recycling: 'I Don't Even Think We Have an Industry,'" *Guardian*, January 9, 2009.

59. Tania Branigan, "From East to West, a Chain Collapses," *Guardian*, January 9, 2009; Christine Oliver, "Recycling in the Credit Crunch," *Guardian*, January 9, 2009; Leo Hickman, "The Truth about Recycling," *Guardian*, February 26, 2009.

60. "UN Programme Aims at Environmentally Sound Disposal of Electronic Waste," *UN News Centre*, November 25, 2005, http://www.un.org/apps/news/story.asp?NewsID=16690&Cr=electronic&Cr1=.

61. Heather Rogers, *Gone Tomorrow: The Hidden Life of Garbage* (New York: New

Press, 2005), 201. Increasingly, the production of electronics has been offshored to India, China, and Taiwan—the same places where electronic waste returns after it has been shipped and consumed in the United States and Europe.

62. Allan Sekula, *Fish Story* (Düsseldorf: Richter Verlag, 1995), 12. See also Marc Levinson, *The Box: How the Shipping Container Made the World Smaller and the World Economy Bigger* (Princeton: Princeton University Press, 2006).

63. Sekula, *Fish Story*, 49.

64. Ibid., 50. Not only is shipping a more protracted space of material movement, but it also requires extensive material resources while contributing significantly to greenhouse gas emissions. See National Oceanic and Atmospheric Administration, "Maritime Shipping Makes Hefty Contribution to Harmful Air Pollution," February 26, 2009, http://www.noaanews.noaa.gov/stories2009/20090226_shipping.html.

65. While North America is currently the largest consumer of electronics, China and India are now experiencing the greatest growth in electronics consumption. In this case, the countries that have traditionally been the recipients of electronic waste from other parts of the world must now deal with their own internal electronic waste problem. See the "Electronic Waste Guide," which notes, "China had the highest growth in number of computer users per capita in the period 1993–2000. It grew a massive 1052%, compared to a world average of 181%" (http://www.ewasteguide.info/economical_facts_and_figures).

66. Basel Action Network and Silicon Valley Toxics Coalition, *Exporting Harm*, 5.

67. Ibid., 18.

68. Zygmunt Bauman, *Wasted Lives: Modernity and Its Outcasts* (Cambridge: Polity, 2004), 60. Bauman notes that electronics are so easily outdated that not even charities will accept them as donations. To process the waste, instead, there emerges "human-waste producing plants. In Guiyu, there are 100,000 of them—men, women and children working for the equivalent of 94p a day."

69. Gavin Lucas, "Disposability and Dispossession in the Twentieth Century," *Journal of Material Culture* 7, no. 1 (2002): 15.

70. Moser addresses this dynamic within recycling: "As potential resource, waste therefore reacquires value and becomes a commodity. It then can be treated as any other product, even traded on the stock exchange, on the condition, however, that it undergo a process of transformation, which sees it from a heterogeneous and degraded object back to the status of basic material" ("Acculturation of Waste," 96).

71. Martin O'Brien, "Rubbish-Power: Towards a Sociology of the Rubbish Society," in *Consuming Cultures*, ed. Jeff Hearn and Sasha Roseneil (Houndsmill, UK: Macmillan, 1999), 270. While the disruption that waste effects within traditional political economies of production and consumption is of greater interest here, O'Brien does expand his study of waste from rubbish economies to histories, literatures, and more in *A Crisis of Waste? Understanding the Rubbish Society* (London: Routledge, 2007).

72. As Hetherington similarly explains, "'Dirt' has to do with the making and unmaking of that process rather than with a thing in itself" ("Secondhandedness," 163). Hetherington draws on discussions of dirt by Lyotard and Serres: see Jean-François Lyotard, *Driftworks* (New York: Semiotext(e), 1984); Michel Serres, *Rome* (Stanford: Stanford University Press, 1991).

73. Karl Marx, *Grundrisse: Foundations of the Critique of Political Economy (Rough Draft)*, trans. Martin Nicolaus (London: Pelican, 1973), 91.

74. Ibid., 93.

75. Ibid.

76. Lucas, "Disposability and Dispossession,"17.

77. Rudi Colloredo-Mansfeld, "Consuming Andean Televisions," *Journal of Material Culture* 8, no. 3 (2003): 275, 283.

78. Lucas, "Disposability and Dispossession," 19.

Chapter 4

Ted Nelson, *Computer Lib/Dream Machines* (Redmond, WA: Tempus Books of Microsoft Press, 1987), 4–5.

Stiegler, *Technics and Time*, 27.

1. While a basic distinction is often made between the museum as a space of exhibition and the archive as a space of storage, I will discuss the museum and archive together in this chapter, as electronic technologies tend to collapse the distance between these two previously distinct entities. For more on the convergence of museum and archive (and library), see Wolfgang Ernst, "Archi(ve)textures of Museology," in *Museums and Memory*, ed. Susan A. Crane (Stanford: Stanford University Press, 2000), 17–34.

2. Straw, "Exhausted Commodities."

3. "In regard to such a perception, one could speak of the increasing concentration (integration) of reality, such that everything past (in its time) can acquire a higher grade of actuality than it had in the moment of its existing. How it marks itself as higher actuality is determined by the image as which and in which it is comprehended. And this dialectical penetration and actualization of former contexts puts the truth of all present action to the test. Or rather, it serves to ignite the explosive materials that are latent in what has been (the authentic figure of which is *fashion*)" (Benjamin, *Arcades Project*, 392).

4. As further explained by Rosalind Krauss, "Benjamin believed that at the birth of a given social form or technological process the utopian dimension was present and, furthermore, that it is precisely at the moment of the obsolescence of that technology that it once more releases this dimension, like the last gleam of a dying star. For obsolescence, the very law of commodity production both frees the outmoded object from the grip of utility and reveals the hollow promise of that law" (*A Voyage on the North Sea: Art in the Age of the Post-Medium Condition* [London: Thames and Hudson, 1999], 41).

5. Benjamin, *The Origin of German Tragic Drama*, 182. See also Buck-Morss, *Dialectics of Seeing*, 160.

6. See Benjamin, *Arcades Project*, 462; Benjamin, *The Origin of German Tragic Drama*, 178.

7. Hetherington, "Secondhandedness," 166. Hetherington writes, "There are few things that institutions like museums and libraries agonize over more than disposal in the sense of deaccessioning." Indeed, in various computing history archives, one finds such an uneasy relation to the fact that the sheer number of preservable electronics well exceeds the holding capacity of the museum. So the inevitable deaccessioning takes place, where objects are given nearly ritualistic burials at the local landfill. This is not a topic most curators are prepared to discuss, but the fact remains that not every object in the history of electronics can possibly be retained for "future posterity."

8. This chapter maintains the standard differentiation between mass or permanent storage, typically on magnetic media, and memory, which refers to the temporary working storage of random-access memory (RAM). In common usage, however, these terms are often used interchangeably.

9. Seitz and Einspruch, *Electronic Genie*, 213–15.

10. The manipulation of time in fact counts as the measure of progress for electronics in multiple ways. As Bolter writes, "The operating time is often the single most important measure of work done by the computer. When a new machine is brought on the market, the first question asked is: how fast are the basic instructions to fetch data from memory, operate upon it, and return the result? A new computer installation, a collection of several processors and storage devices, measures its productivity in terms of *throughput*, that is, how many programs can be run in a fixed period of time. Conversely, the programmer grades the success of his solution to a problem by the speed of execution of his program" (*Turing's Man*, 109–10).

11. Jean François Lyotard, *The Inhuman: Reflections on Time*, trans. Geoffrey Bennington and Rachel Bowlby (Cambridge: Polity, 1991), 64.

12. Arthur C. Clarke, *Profiles of the Future: An Inquiry into the Limits of the Possible* (London: Indigo, 1999), 198.

13. Ibid.

14. Hansen's discussion of voluntary and involuntary memory as read through Benjamin may be a helpful addition to this statement: "For Benjamin, therefore, the disjunction demarcates two antithetical types of experience: one centered around a reflective, psychic subject whose powers have been markedly diminished with the advent of modernity (*Erfahrung);* another around a corporeal agency sensitive to the inhuman rhythms of the mechanosphere (*Erlebnis*). Since voluntary memory takes its standard directly from the rhythm of external duration, of the commodity world itself, its predominance in the modern world yields a fundamental deterritorialization of the traditional humanist, cognitive subject" (Hansen, *Embodying Technesis*, 243).

15. Bush, "As We May Think," 101–8.

16. Jim Gemmell, Gordon Bell, and Roger Lueder, "MyLifeBits: A Personal Database for Everything," Microsoft Research Technical Report, MSR-TR-2006-23 (San Francisco: Microsoft Bay Area Research Center, 2006), http://research.microsoft.com/apps/pubs/default.aspx?id=64157.

17. Ibid.

18. Ibid. See also Gordon Bell and Jim Gemmell, *Total Recall: How the E-Memory Revolution Will Change Everything* (New York: Dutton, 2009).

19. As Friedrich Kittler argues, "For the very first time in media history, data that are stored or transferred are already computable for that very reason" ("Museums on the Digital Frontier," in *The End(s) of the Museum*, ed. Alexander Garcia Düttmann et al. [Barcelona: Fundació Antoni Tàpies, 1996], 71).

20. Alan Turing, "Intelligent Machinery," in *Mechanical Intelligence*, ed. D. C. Ince (1948; repr., London: North Holland, 1992), 107–28.

21. Kittler, "Museums on the Digital Frontier," 75.

22. Corzo asks, "Is it possible that the clay tablets of Tell Brak will last longer than our current high-powered, ultra-sophisticated technology?" (Miguel Angel Corzo, ed., *Mortality/Immortality? The Legacy of 20th-Century Art* [Los Angeles: Getty Conservation Institute, 1999], xvii).

23. Gemmell, Bell, and Lueder, "MyLifeBits."

24. Indeed, MyLifeBits assumes such an underlying condition of transience, which presents cause for one's personal database to grow. This is due to the fact that "external" data, such as Web sites, are so transient that it is necessary to store this desired data on one's own system in order to ensure continuing access.

25. In Chris Marker's film *Sans Soleil* (1982), the narrator predicts, "The New Bible will be of Magnetic Memory, and will have to rerecord itself constantly just to remember itself." This is a situation of "total recall" and "total amnesia," of a memory that is as extended as it is volatile.

26. Friedrich Kittler writes, "Emulation would seem to be the answer to the oft-repeated paradox that the computer as a medium can archive all other media except itself. In his essay *Trancemedia: from Simulation to Emulation,* Arjen Mulder suggests that emulation is the only means at the computer's disposal to secure and access its own history. Emulation of all earlier hardware plus software is the only way in which computer history can be written using a computer" ("The Exhibition as Emulator," trans. James Boekbinder, text commissioned for the 2000 InfoArcadia exhibition, http://www.mediamatic.net/article-8740-en.html).

27. Molotch, *Where Stuff Comes From,* 2–3.

28. Bruce Sterling, "Built on Digital Sand: A Media Archaeologist Digs the Lonely Shores of Binary Obsolescence," in "Ghost: Archive, Evolution, Entropy," *Horizon Zero* 18 (2004), http://www.horizonzero.ca/textsite/ghost.php?is= 18&file =4&tlang=0.

29. Ibid.

30. Ibid.

31. Geert Lovink, "Archive Rumblings: Interview with German Media Archaeologist Wolfgang Ernst," Nettime, February 25, 2003, http://www.nettime.org/Lists-Archives/nettime-l-0302/msg00132.html.

32. Ibid.

33. This concept is expressed in varying ways by authors as far-ranging as Stiegler, Hansen, Lyotard, and Sterling. For a more extended discussion, see especially Stiegler's *Technics and Time.* In this work, Stiegler writes, "There is a *historicity* to the technical object that makes its descriptions as a mere hump of inert matter impossible. This inorganic matter organizes *itself.* In organizing itself, it becomes indivisible and conquers a quasi-ipseity from which its dynamic proceeds absolutely: the history of this becoming-organic is not that of the humans who 'made' the object" (71).

34. Bruce Sterling, *Shaping Things* (Cambridge, MA: MIT Press, 2005), 58.

35. Franz Alt, *Evaluation of Automatic Computing Machines* (Washington, DC: National Bureau of Standards, 1951).

36. Ceruzzi, *History of Modern Computing,* ix.

37. Ibid.

38. A number of scholars write on the multiple implications of this term. See, for example, Jonathan Sterne, "Out with the Trash: On the Future of New Media," in *Residual Media,* ed. Charles Acland (Minneapolis: University of Minnesota Press, 2007), 16–31; Wendy Hui Kyong Chun and Thomas Keenan, eds., *New Media, Old Media* (New York: Routledge, 2006); Lisa Gitelman, *Always Already New: Media, History, and the Data of Culture* (Cambridge, MA: MIT Press, 2006). Writing in what seems to be a Benjaminian register, Paul Rabinow suggests the "contemporary" emerges not through the elimination of the new, but through the shaping of distinc-

tive relations between "older and newer elements." See *Marking Time* (Princeton: Princeton University Press, 2007).

39. Packard identifies three modes of obsolescence that range from function to fashion, including "obsolescence of function," "obsolescence of quality," and "obsolescence of desirability" (*Waste Makers*, 55).

40. Toffler, *Future Shock*, 62–63.

41. Ibid., 62.

42. Industrial designer Brooke Stevens is often credited with popularizing the term *obsolescence*. Slade suggests that Stevens was actually preceded by other promoters, who also championed obsolescence as a stimulant to economic activity. Most notably, Bernard London proposed death dating for objects, which would have predetermined life spans and would have to be returned to manufacturers upon expiration. See Slade, *Made to Break*, 73–75.

43. The Long Now Foundation suggests that "by constantly accelerating its own capabilities (making faster, cheaper, sharper, tools that make even faster, cheaper, sharper tools), the technology is just as constantly self-obsolescing. The great creator becomes the great eraser." Yet this erasure is never complete, because "behind every hot new working computer is a trail of bodies of extinct computers, extinct storage media, extinct applications, extinct files." See The Long Now Foundation, http://www.longnow.org; Stewart Brand, *The Clock of the Long Now: Time and Responsibility* (New York: Basic Books, 1999).

44. See also Bill Joy, "Why the Future Doesn't Need Us," *Wired News* 8, no. 4 (April 2000). The history of human obsolescence could constitute a book in itself: both nineteenth- and twentieth-century versions of automation have given rise to distinct forms of human obsolescence. I would title this book *The Post-Luddite Chronicles*, as a way to engage with this knotted intersection and humanist dilemma where the human passes out of relevance only to return as an imagined site of resistance in relation to technologies.

45. Even going beyond the typical analysis of actor-network theory, which places humans and nonhumans on a relatively flattened field of influence, the obsolescence of the human suggests a condition more similar to that identified by Serres, where we become subjected to our objects, with the result that the very possibility of sharply delineated causal relations between subjects and objects crumbles. See Michel Serres, *The Natural Contract*, trans. Elizabeth MacArthur and William Paulson (Ann Arbor: University of Michigan Press, 1995).

46. Sterling, *Shaping Things*, 59.

47. Evan Watkins, *Throwaways: Work Culture and Consumer Education* (Stanford: Stanford University Press, 1993), 26.

48. Ibid., 32. Straw similarly writes, in "Exhausted Commodities," that obsolete objects do not "simply disappear"; rather, they "persist and circulate through the commercial markets of contemporary life."

49. Watkins, *Throwaways*, 19.

50. Moore, "Cramming More Components onto Integrated Circuits."

51. Ceruzzi, *History of Modern Computing*, 7.

52. On this note, Ceruzzi asks, "Yet who would deny that computing technology has been anything short of revolutionary? A simple measure of the computing abilities of modern machines reveals a rate of advance not matched by other technologies, ancient or modern" (ibid., 3).

53. According to Ceruzzi, the *pervasiveness* of computers also attests to the far-reaching impacts of this revolution. In this respect, he suggests that computers are among the most significant technological devices, more revolutionary than the washing machine (ibid., 3). But the washing machine is also now a computer. As an appliance, it, too, has become a device for the distribution of microchips.

54. Donald MacKenzie, *Knowing Machines: Essays on Technical Change* (Cambridge, MA: MIT Press, 1998), 8.

55. Ibid.

56. Gordon Moore, interview, *Silicon Genesis*. http://silicongenesis.stanford.edu/complete_listing.html.

57. MacKenzie, *Knowing Machines*, 56.

58. MacKenzie writes, "The prophecy of a specific rate of increase has thus been self-fulfilling. It has clearly served as an incentive to technological ambition; it has also, albeit less obviously, served to limit such ambition" (ibid.). As Michel Callon underscores, self-fulfilling prophecies are never simple scripts that are directly implemented without consequence. Indeed, the failure—as much as the success—of these implementations reveals the complex material and textual worlds in which these prophecies unfold. See Michel Callon, "What Does It Mean to Say That Economics Is Performative?" 323–24.

59. MacKenzie, *Knowing Machines*, 56.

60. Toffler, *Future Shock*, 53.

61. It may be helpful here to remember Appadurai's discussion of turnover, where he writes, "From the point of view of demand, the critical difference between modern, capitalist societies and those based on simpler forms of technology and labor is *not* that we have a thoroughly commoditized economy whereas theirs is one in which subsistence is dominant and commodity exchange has made only limited inroads, but rather that the consumption demands of persons in our own society are regulated by high-turnover criteria of 'appropriateness' (fashion), in contrast to the less frequent shifts in more directly regulated sumptuary or customary systems. In both cases, however, demand is a socially regulated and generated impulse, not an artifact of individual whims or needs." As he suggests, it is exactly this rate of turnover, not consumption or commodification per se, that distinguishes economies. See Appadurai, "Introduction," 32.

62. Toffler further writes in the same section, "Sophisticates in the fad business prepare in advance for shorter and shorter product life cycles. Thus, there is in San Gabriel, California, a company entitled, with a kind of cornball relish, Wham-O Manufacturing Company. Wham-O specializes in fad products, having introduced the hula hoop in the fifties and the so-called Super-Ball more recently. . . . Wham-O and other companies like it, however, are not disconcerted when sudden death overtakes their product; they anticipate it. They are specialists in the design and manufacture of 'temporary' products" (*Future Shock*, 66).

63. Mike Featherstone, "Archiving Cultures," *British Journal of Sociology* 51, no. 1 (January/March 2000): 170.

64. Ernst also suggests, "The electronic archive no longer emphatically differentiates between memory and waste. There is no technical distinction between sense and nonsense" (Wolfgang Ernst, "Agencies of Cultural Feedback: The Infrastructure of Memory," in Neville and Villeneuve, *Waste-Site Stories*, 113).

65. Ibid., 115.

66. Ibid., 116.

67. Jacques Derrida, *Archive Fever: A Freudian Impression,* trans. Eric Prenowitz (Chicago: University of Chicago Press, 1996), 16.

68. Ibid., 9.

69. Both the "imperative to remember," as Ernst writes, and the need for "intentional forgetting," as the MyLifeBits researchers indicate, need to be programmed into electronic memory technologies. In both respects, electronic circuits are directed to emulate our own memory operations but inevitably perform in much different ways.

70. See David Grattan and R. Scott Williams, "From '91' to '42': Questions of Conservation for Modern Materials," in Corzo, *Mortality/Immortality?* 67–74.

71. Buchli, *Material Culture Reader,* 14.

72. Ibid., 12. As Buchli explains, "material culture studies as part of a sacrificial economy has historically occurred within a framework of social purpose, which required the constitution of such super-material objects—material culture—to facilitate these goals whether industrial progress, social revolution or critical consciousness."

73. As design historian Adrian Forty writes, "The Western tradition of memory since the Renaissance has been founded upon an assumption that material objects, whether natural or artificial, can act as the analogues of human memory. It has been generally taken for granted that memories, formed in the mind, can be transferred to solid material objects, which can come to stand for memories and, by virtue of their durability, either prolong or preserve them indefinitely beyond their purely mental existence." However, "it is clear that the relationship between objects and memory is less straightforward than Western thinking has been in the habit of assuming. We cannot take it for granted that artifacts act as the agents of collective memory, nor can they be relied upon to prolong it." See Adrian Forty and Susanne Küchler, eds., *The Art of Forgetting* (Oxford: Berg, 1999), 2, 7.

74. Ernst, "Archi(ve)textures of Museology," 25–28. As Ernst suggests in this respect, "Maybe the task of the museum today is to reflect on the contemporary loss of substance in objects; a contributor to this dematerialization, however, is the museum itself."

75. See the Museum of E-Failure, http://www.disobey.com/ghostsites/.

76. See the Internet Archive, http://www.archive.org. Elsewhere, Ernst suggests, "The Internet has turned the notion of the archive into a metaphor for *data retrieval*" ("Agencies of Cultural Feedback," 117).

77. Featherstone, "Archiving Cultures," 179.

78. Jens Schröter, "Archive—Post/photographic," Media Art Net, http://www.medienkunstnetz.de/themes/photo_byte/archive_post_photographic/.

79. Schröter further writes (in a futuristic, if progressive, tone), "In this, the issue is not a nostalgic return to the safe refuge of the museum—which is impossible as it is—rather, the point is to keep the discussion on the archive of the future in motion, that is, of working on alternative models to the archive" (ibid.).

80. See Hayles, *My Mother Was a Computer,* 89–104. One striking project that engages with emulation through correspondences is the Variable Media Network, a collaboration between the Guggenheim Museum in New York and the Daniel Langlois Foundation for Art, Science, and Technology. This project preserves the ephemeral media and the obsolescence of electronic technology in artists' projects by migrating projects across media. See http://www.variablemedia.net.

81. Kittler, "Exhibition as Emulator."

82. In fact, such a description might align with Benjamin's description of a preferred mode of storytelling, which consists of the "slow piling one on top of the other of thin, transparent layers which constitutes the most appropriate picture of the way in which the perfect narrative is revealed through the layers of a variety of retellings." See Walter Benjamin, "The Storyteller: Reflections on the Works of Nikolai Leskov," in *Illuminations*, 93.

83. Kittler, "Museums on the Digital Frontier," 78.

84. Ernst, "Agencies of Cultural Feedback," 116.

85. Moser, "Acculturation of Waste," 102.

86. Ernst draws out the possibilities of what happens at the intersection of memory and waste: "The digital archive may merely aim at storing information; the energetic texture of memory, though, is based on counter- and dis-information, which amounts to a memory of waste" ("Agencies of Cultural Feedback," 118).

Chapter 5

Philip K. Dick, *In Milton Lumky Territory* (1985; repr., London: Paladin, 1987), 61.

1. Basel Action Network and Silicon Valley Toxics Coalition, *Exporting Harm.*

2. Rathje and Murphy, *Rubbish!* 117

3. In this respect, Neville and Villeneuve write, "We observe, moreover, that waste is defined by its own resistant materiality, by what we are tempted to call its 'material memory'" ("Introduction," *Waste-Site Stories*, 7).

4. Ilya Kabakov, "The Man Who Never Threw Anything Away," in *The Archive*, ed. Charles Merewether (London: Whitechapel Gallery; Cambridge, MA: MIT Press, 2006), 37.

5. Michael Shanks, *Experiencing the Past: On the Character of Archaeology* (London: Routledge, 1992), 75.

6. Brown, "Thing Theory," 4–8. Brown writes more extensively on the concept of breakdown, noting, "We begin to confront the thingness of objects when they stop working for us: when the drill breaks, when the car stalls, when the windows get filthy, when their flow within the circuits of production and distribution, consumption and exhibition, has been arrested, however momentarily."

7. Benjamin draws on the urban poetry of Charles Baudelaire to describe this particular form of salvage, or ragpicking. Citing Baudelaire, he writes, "'Here we have a man whose job it is to gather the day's refuse in the capital. Everything that the big city has thrown away, everything it has lost, everything it has scorned, everything it has crushed underfoot he catalogues and collects. He collates the annals of intemperance, the capharnaum of waste. He sorts things out and selects judiciously: he collects like a miser guarding a treasure, refuse which will assume the shape of useful or gratifying objects between the jaws of the goddess of Industry.' This description is one extended metaphor for the poetic method, as Baudelaire practiced it. Ragpicker and poet: both are concerned with refuse." See Walter Benjamin, "The Paris of the Second Empire in Baudelaire," in *Selected Writings*, vol. 4, *1938–1940*, ed. Howard Eiland and Michael W. Jennings, trans. Edmund Jephcott et al. (Cambridge: Harvard University Press, Belknap, 2002), 48; Ursula Marx et al., eds., "Ragpicking: *The Arcades Project*," in *Walter Benjamin's Archive: Images, Texts, Signs*, trans. Esther Leslie (London: Verso, 2007), 251–65.

8. For an urban narrative approach to salvaging and the possible imaginaries to which this gives rise, see Jeff Ferrell, *Empire of Scrounge: Inside the Urban Underground of Dumpster Diving, Trash Picking, and Street Scavenging* (New York: New York University Press, 2006).

9. Joost van Loon and Ida Sabelis, "Recycling Time: The Temporal Complexity of Waste Management," *Time & Society* 6, no. 2 (June 1997): 293. As Van Loon and Sabelis write, "The temporal paradoxes of waste are hidden in an idealistic, technocratic and future-oriented ethos that fails to take into account the continuity of its own implications in the present, that is, the time-scales of waste itself."

10. Thompson, *Rubbish Theory*, 9.

11. Kevin Lynch, *Wasting Away* (San Francisco: Sierra Club Books, 1990), 69.

12. International Association of Electronics Recyclers, *IAER Electronics Recycling Industry Report*, 7. For a more extended discussion of the cultural possibilities of repair (from a relatively Heideggerian perspective), see Stephen Graham and Nigel Thrift, "Out of Order: Understanding Repair and Maintenance," *Theory, Culture & Society* 24, no. 3 (2007): 1–25.

13. National Safety Council, *Electronic Product Recovery and Recycling Baseline Report: Recycling of Selected Electronic Products in the United States* (Washington, DC, 1999), 36.

14. International Association of Electronics Recyclers, *IAER Electronics Recycling Industry Report*, 32.

15. Lynch, *Wasting Away*, 61.

16. Peter J. M. Nas and Rivke Jaffe, "Informal Waste Management: Shifting the Focus from Problem to Potential," *Environment, Development, and Sustainability* 6 (2004): 339.

17. Ravi Agarwal and Kishore Wankhade, "Hi-Tech Heaps, Forsaken Lives: E-waste in Delhi," in Smith, Sonnenfeld, and Pellow, *Challenging the Chip*, 237. Several organizations covering the topic of electronic waste workers document the working conditions and practices involved with the salvaging of electronics. The Basel Action Network and Toxics Link are two such organizations that have documented processes of electronic waste sifting in developing countries. See also Greenpeace, "Recycling of Electronic Wastes in China & India: Workplace & Environmental Contamination," August 2005, http://www.greenpeace.to/publications/electronic_waste_recycling_appendix.pdf.

18. Ibid., 237.

19. Ibid., 239.

20. Nas and Jaffe, "Informal Waste Management," 343.

21. See Karen Tranberg Hansen, *Salaula: The World of Secondhand Clothing and Zambia* (Chicago: University of Chicago Press, 2000), for a detailed account of the salvage practices in relation to clothing donated to charity and shipped to places such as Africa. These salvage practices extend to more than simply receiving items, as Hansen documents, also involving transforming them in entirely creative ways.

22. John Frow, "A Pebble, a Camera, a Man Who Turns into a Telegraph Pole," *Critical Inquiry*, 28, no. 1, Things (Autumn 2001): 284.

23. Colloredo-Mansfeld, "Introduction," 250.

24. Taussig, *Mimesis and Alterity*, 232. Here Taussig is drawing on the atmosphere released from outmoded objects as discussed by Benjamin in his essay "Surrealism."

25. Ibid.

26. Ibid.

27. Agarwal and Wankhade, "Hi-Tech Heaps, Forsaken Lives," 234–46.

28. Hawkins writes on the "cultural economy" of recycling and explores the extent to which, from the 1960s onward, recycling has emerged out of a sense of "nature in crisis," as it was transferred to households and consumers as the site of management. See Hawkins, *Ethics of Waste,* 99–104. Within the context of this focus on consumers and recycling, it is interesting to note just how little of the waste stream is composed of end-consumer waste. Even accounting for how much general waste industry produces (and potentially recycles) in comparison to consumers is a difficult task, moreover. Molotch writes, "What consumers have been less able to affect is the recycling of producers' waste, a major problem given that, according to the U.S. Environmental Protection Agency (EPA), *only about 2 percent* of all waste comes from households, offices, institutions, and retail in the first place—'Municipal Solid Waste,' in the official terminology. Given that hazardous waste makes up about 6 percent of the total, much of it indeed controlled in various ways across the jurisdictions, 92 percent of U.S. waste is unaccounted for; there are almost no data on what makes it up. Most is apparently disposed of on site, beyond prying eyes" (*Where Stuff Comes From,* 235). Matthew Gandy similarly writes, "Municipal waste accounts for only a relatively small fraction of total global waste production, the main sources being from agriculture, industry and mining" (*Recycling and the Politics of Urban Waste* [New York: St. Martin's, 1994], 4).

29. Loon and Sabelis, "Recycling Time," 294. As already mentioned, these viable future markets have recently been put into question with the temporary lapse during the 2007 financial crisis with China accepting recyclables, volatility in material prices, and realignments of material markets.

30. Ibid., 295.

31. Ibid., 302.

32. See, for instance, Shred Tech's approach to electronic waste handling, described at http://www.shred-tech.com/electronic.html.

33. Basel Action Network and Silicon Valley Toxics Coalition, *Exporting Harm,* 17.

34. Carolyn Steedman, *Dust: The Archive and Cultural History* (Manchester: Manchester University Press, 2001), 163.

35. Ibid., 164.

36. Ibid.

37. Basel Action Network and Silicon Valley Toxics Coalition, *Exporting Harm,* 6.

38. Ibid., 6. As these authors argue, "The assumption too, is that recycling is always better than landfilling. This is not the case when the recycling results in toxic worker exposures, and the open dumping or burning of toxic residues and wastes that we have witnessed in Asia. While the recycling of hazardous materials anywhere creates a serious pollution challenge, it can be a disastrous one in an area of the world where the knowledge of, and infrastructure to deal with hazards and waste is almost non-existent. With the cautionary note above, it is nevertheless estimated that in 1998, 11% of computers were being recycled (including those sent for export)."

39. Ibid., 7. On the quantity of electronic waste that issues for the dump, these authors write, "According to the EPA, in 1997 more than 3.2 million tons of E-waste ended up in U.S. landfills. It is thought that most households and small businesses

that dispose rather than store their obsolete electronic components send their material to landfills or incinerators rather than take them to recyclers."

40. Ibid., 21–22.

41. Lynch, *Wasting Away,* 54.

42. Ibid., 191.

43. Mira Engler, *Designing America's Waste Landscapes* (Baltimore: Johns Hopkins University Press, 2004), 112–13. See also Noel van Dooren, "Never Again Will the Heap Lie in Peace," in *Tales of the Tip,* ed. Chris Driessen and Hiedi van Mierlo (Amsterdam: Fundament Foundation, 1999), 100–105.

44. O'Brien, "Rubbish-Power," 268. This is the "contrariness of waste," as O'Brien writes, "its capacity to be its own opposite, to have no apparent value and yet potentially to be valuable."

45. Loon and Sabelis, "Recycling Time," 299.

46. Basel Action Network and Silicon Valley Toxics Coalition, *Exporting Harm,* 7.

47. Ibid., 27. As Jim Puckett writes, "Indeed, hazardous waste recycling is a dangerous practice anywhere on earth, even in rich, developed countries. To this day, BAN's onsite visits to electronic waste processors in the United States reveal a very serious lack of concern and knowledge about the impacts of such hazards as brominated flame retardants and beryllium in fumes and dusts, as well as toxic cadmium and rare-earth, metal-laced phosphors released by broken cathode ray tubes (CRTs). And smelters, the final and weakest environmental link in any metals recycling operation, have been historically notorious in Europe and North America as major point-source polluters" ("High-Tech's Dirty Little Secret: The Economics and Ethics of the Electronic Waste Trade," in Smith, Sonnenfeld, and Pellow, *Challenging the Chip,* 229).

48. For additional information on substances classified as harmful or hazardous, see CERCLA, "Priority List of Hazardous Substances," http://www.atsdr.cdc.gov/cercla/index.asp.

49. Marilyn Strathern, *Property, Substance, and Effect: Anthropological Essays on Persons and Things* (London: Athlone, 1999), 61.

50. Loon and Sabelis, "Recycling Time," 297. As these authors write, "There is little doubt that time concepts are of eminent importance when discussing environmental matters. If recycling based on a cyclical-linear time is our ideal, we might falsely believe that damage can be undone."

51. Buchli, *Material Culture Reader,* 17.

52. Scanlan suggests that such decay opens up orders of time that we too often prefer to ignore. "Deteriorating matter," he writes, "embodies a time that exists beyond our rational time: in this shadow world, time is always running matter down, breaking things into pieces, or removing the sheen of a glossy surface and, therefore, the principal methods of dealing with material waste throughout most of human history—dumping, burning, recycling, reducing the use of virgin materials—are simply ways of ensuring that this fact does not intrude too far into everyday experience. In contemporary society the increase in the volume of consumer products may force a strict reorganization of time, in so far as the situation can be met by adherence to a regime that ensures the removal of these objects before they decay" (*On Garbage,* 34).

53. Rem Koolhaas, "Wasteland—Dump Space: Freedom from Order," *Wired* 11, no. 6 (June 2003).

54. Douglas, *Purity and Danger,* 198.

55. Ibid., 196.

56. On Benjamin's interest in undoing the insistence on progress, Peter Osborne writes, "The avant-garde is not that which is most historically advanced because it has the most history behind it (the angel), but that which, in the flash of the dialectical image, disrupts the continuity of 'progress,' and is thus able (like the child) to 'discover the new anew.' Benjamin's philosophy of history is a struggle to wrestle the idea of the 'new,' essential to any concept of the avant-garde, away from the ideology of 'progress'" ("Small-Scale Victories, Large-Scale Defeats: Walter Benjamin's Politics of Time," in *Walter Benjamin's Philosophy: Destruction and Experience,* ed. Andrew Benjamin and Peter Osborne [Manchester: Clinamen, 2000], 88).

Conclusion

Michel Serres, *Genesis,* trans. Genevieve James and James Nielson (Ann Arbor: University of Michigan Press, 1995), 91.

Serres, *Parasite,* 68.

1. Lynch, *Wasting Away,* 6.
2. Ibid., 6–8.
3. Ibid., 10.
4. For example, see Murray, *Zero Waste.*
5. "Waste management," as Loon and Sabelis write, not only entails "the commodification of excess productivity into renewable resources," it also "becomes a euphemism for the displacement and misplacement of the unwanted and unmanageable consequences of modernization" ("Recycling Time," 292).
6. Ibid., 296. Loon and Sabelis further write, there emerges "a new, higher moral union," where "the environment enters the capitalist system through commodification, while in return capitalism is further mythified as 'natural.'" There are numerous texts and projects that have been developed within the natural-capital nexus. For one example relevant to design, see William McDonough and Michael Braungart, *Cradle to Cradle: Remaking the Way We Make Things* (New York: North Point Press, 2002).
7. Moser similarly writes, "Waste is permanent and unavoidable, for there is no system—whether biological, technical, social, or historical—that does not produce remnants, remains, scraps, leftovers, that does not leave certain parts to decay, that does not secrete or reject. Anything in a system can become waste" ("Acculturation of Waste," 102).
8. Serres, *Parasite,* 12–13. I discuss the notions of "systems" and "harmony" at greater length in Jennifer Gabrys, "Sink: The Dirt of Systems," *Environment and Planning D: Society and Space* 27, no. 4 (2009): 666–81.
9. Hawkins, *Ethics of Waste,* 122.
10. For an example of this sort of return of waste to productive mechanisms, consider the proposal Judd H. Alexander makes in his book *In Defense of Garbage,* where he suggests that garbage is a good thing because it offers a way to fill up all the holes that result from mining and other forms of resource extraction. He calculates that in the United States, more than twenty-three times the amount of land is removed through resource extraction than is filled. Clearly, the solution is to fill all the holes with garbage. In Alexander's strange mathematics, the distinc-

tion between raw material, commodity, and waste collapses. Materials seem only to have wanted to leave the ground to return as trash. But the transformation that these materials undergo and the remainders and irreversible effects they generate comprise a critical distinction and an overlooked area of investigation. See Judd H. Alexander, *In Defense of Garbage* (Westport, CT: Greenwood, 1993).

11. Loon and Sabelis, "Recycling Time," 296. Addressing the mythic solution engineered by waste management, these authors write, "Progress is simply conceived of in terms of a differential discount on the future. This means that while progress is accounted for in the present, it entails a reduction of the complexity of futures that springs from the indeterminacy of the difference between the actual and the possible."

12. Ibid., 297. These authors make the statement, "Markets cannot resolve ecological issues because they are not geared toward providing infrastructures for dealing with negative utility (excess). The win-win economics of ecological marketing ('it is possible to capitalize on waste') will not pay off if the hidden costs are taken into account" (302).

13. The difficulty of even agreeing on what constitutes waste and what the severity of certain types of waste consists of has meant that hazardous waste may not be acknowledged as such and will be improperly handled. This is certainly true with electronics, where the apparent inertness of these devices conceals the hazards that lie within the machines. Electronic waste and other forms of waste often waver in such spaces of indeterminable status. As *The Guardian* reports, "One of the great challenges of our time is to collectively agree on what is waste and what are second-hand products—this question extends to end-of-life ships as much as to electronic goods" (Hilary Osborne, "Rich Nations Accused of Dumping E-Waste on Africa," *Guardian*, November 27, 2006).

14. Hawkins, *Ethics of Waste*, 122.

15. Ibid.

16. Loon and Sabelis, "Recycling Time," 295, 298.

17. Ibid., 303.

18. Hawkins, *Ethics of Waste*, 81.

19. See Electronic Product Environmental Assessment Tool, http://www.epeat.net; Silicon Valley Toxics Coalition, "Green Chemistry," http://www.svtc.org/site/PageServer?pagename=svtc_green_chemistry; Green Electronics Council with the National Center for Electronics Recycling and Resource Recycling, "Closing the Loop: Electronics Design to Enhance Reuse/Recycling Value," January 2009, http://www.greenelectronicscouncil.org/documents/0000/0007/Design_for_End_of_Life_Final_Report_090208.pdf; E-Stewards Initiative, http://www.e-stewards.org/; Solving the E-waste Problem (StEP), "Annual Report, 2009," http://www.step-initiative.org/pdf/annual-report/Annual_Report_2009.pdf; Sibylle D. Frey, David J. Harrison, and Eric H. Billett, "Ecological Footprint Analysis Applied to Mobile Phones," *Journal of Industrial Ecology* 10, nos. 1–2 (2006): 199–216; C. Kieren Mayers, Chris M. France, and Sarah J. Cowell, "Extended Producer Responsibility for Waste Electronics," *Journal of Industrial Ecology* 9, no. 3 (2005): 169–89.

20. Molotch, *Where Stuff Comes From*, 245–46.

21. Greenpeace, "Guide to Greener Electronics," December 2008, http://www.greenpeace.org/rankingguide. Together with improving electronic design through the use of fewer toxic materials, this report proposes extended producer responsibil-

ity, or EPR, as an important strategy in the take-back and recycling of electronics.

22. These initiatives have also emerged in response to the RoHS and WEEE directives in Europe, which have begun to inform the manufacture and disposal of electronics (as previously discussed in this study). For a review of this program, see United Nations University, "Review of Directive 2002/96 on Waste Electrical and Electronic Equipment (WEEE)" (Tokyo: UNU, 2008). See also "Dell's 2008 Corporate Responsibility Summary Report" (2009), http://www.dell.com/sustainability report, for an example of a current approach to electronics that focuses on product life cycle, hazardous substances, and product take-back. U.S. responses have often occurred at the level of state legislation, but for a more comprehensive, if voluntary, approach developed by the EPA, see "Responsible Recycling ('R2') Practices for Use in Accredited Certification Programs for Electronics Recyclers," October 30, 2008, http://www.epa.gov/waste/conserve/materials/ecycling/r2practices.htm; see also Basel Action Network, "Detailed Critique of Problems with R2 Standard," November 2008, http://e-stewards.org/wp-content/uploads/2010/02/Detailed_R2_Analysis.pdf.

23. When I first made the proposal for a "green machines" handbook in 2006, there were relatively few examples of green computing in circulation. As this study is completed, an increasing number of projects are developing in this area, including the recent Bill Tomlinson, *Greening through IT: Information Technology for Environmental Sustainability* (Cambridge, MA: MIT Press, 2010).

24. "Compostable Keyboard," as documented in Alastair Fuad-Luke, *The Eco-Design Handbook* (London: Thames and Hudson, 2005); Joseph Chiodo, "Active Disassembly," http://www.activedisassembly.com/index2.html.

25. On these few (among many) examples of reworking the material form of electronics, see "Cool Light Leads to Greener Chips," *BBC News*, June 30, 2006, http://news.bbc.co.uk/1/hi/technology/5128762.stm; "Cardboard PC Case by Lupo," October 21, 2005, http://www.ubergizmo.com/15/archives/2005/10/cardboard_pc_ca.html.

26. Alternative materials and reduced energy consumption are two areas of considerable attention within design projects. For example, see Core 77, "Greener Gadgets Design Competition," http://www.core77.com/competitions/GreenerGadgets. Given concerns over energy use and climate change, increasing attention is now being drawn to the amount of energy that electronic technologies require—not just to power the devices themselves, but also to power the extensive servers, networks, and interlocking systems that allow these devices to communicate. See Bobbie Johnson, "Google's Power-Hungry Data Centres," *Guardian*, May 3, 2009; Richard Wray, "Spam 'Uses as Much Power as 2.1M Homes,'" *Guardian*, April 15, 2009.

27. The project "How Stuff Is Made," conducted by design students and academics, documents the resources, manufacturing processes, and labor and environmental impact of contemporary goods (see http://www.howstuffismade.org). The United Nations Environment Programme has also recently produced a document that focuses on the social aspects to life-cycle analyses. See United Nations Environment Programme, "Guidelines for Social Life Cycle Assessment of Products," DTI/1164/PA (2009), http://www.unep.org./pdf/DTIE_PDFS/DTIx1164xPA_guidelines_sLCA.pdf.

28. Ed van Hinte, *Eternally Yours: Visions on Product Endurance* (Rotterdam: 010 Publishers, 1997), 27. Sterling similarly projects a relatively friction-free future for technologies. He imagines one speculative version of technology that will "eventually rot and go away by itself." This completely biodegradable and "auto-recycling" technology will, when it breaks down, give rise to new "complicated forests, grasslands and coral reefs." But this technology will not require "natural materials"; rather, it will sprout up in a "room-temperature industrial assembly without toxins." See Sterling, *Shaping Things,* 143.

29. For examples of "trash-tracking" projects, see Eric Paulos and Tom Jenkins, "Urban Probes: Encountering Our Emerging Urban Atmospheres," *Proceedings of the SIGCHI Conference on Human Factors in Computing Systems,* April 2–7, 2005. (Portland, Oregon); Trash Track, http://senseable.mit.edu/trashtrack/; Valerie Thomas, "Radio-Frequency Identification: Environmental Applications" (white paper, Foresight in Governance Project, Woodrow Wilson International Center for Scholars, Washington, DC, 2008).

30. Examples of these projects include Natalie Jeremijenko and Proboscis's feral robotic dogs, http://www.nyu.edu/projects/xdesign/feralrobots/. In a related way, Dunne explores how electronics constitute "post-optimal objects," and he seeks to capture the "para-functionality" of electronics in order to consider how these objects may become critical devices and "provide new types of aesthetic experience." See Dunne, *Hertzian Tales,* 12–14.

31. See Jonah Brucker-Cohen, "Scrapyard Challenge Workshops," http://infamia1.infamia.com/coin-operated.com/; Benjamin Gaulon, "Recyclism," http://www.recyclism.com/.

32. See the call for CHI 2010, Jina Huh et al., "Workshop on Examining Appropriation, Re-use, and Maintenance for Sustainability," http://jinah.people.si.umich.edu//chi2010/reuse.html.

33. "Sustainable HCI" approaches range from the informational to the artistic and from the interventionist to the persuasive. See Carl DiSalvo, Kirsten Boehner, Nicholas A. Knouf, and Phoebe Sengers, "Nourishing the Ground for Sustainable HCI: Considerations from Ecologically Engaged Art," *Proceedings of the CHI Conference on Human Factors in Computing Systems,* April 4–9, 2009 (Boston), 385–94; Marcus Froth, Eric Paulos, Christine Satchell, and Paul Dourish, "Pervasive Computing and Environmental Sustainability: Two Conference Workshops," *IEEE CS* 8, no. 1 (January–March 2009): 78–81.

34. Frow, "A Pebble, a Camera, a Man Who Turns into a Telegraph Pole," 273–74.

35. Felix Guattari's discussion of "three ecologies," spanning from the individual to the sociocultural and environmental, is a relevant reference for addressing the multiple versions of ecologies that inform environmental issues. See Felix Guattari, *The Three Ecologies,* trans. Ian Pindar and Paul Sutton (London: Athlone, 2000).

36. Zehle's proposal is based on James Boyle's article "A Politics of Intellectual Property: Environmentalism for the Net?" 1997, http://www.law.duke.edu/boylesite/Intprop.htm. Where Boyle proposes environmentalism as an analogy for how to negotiate the digital commons of the Internet, Zehle suggests we take this environmentalism more literally into the realm of digital material effects. See Soenke Zehle, "Environmentalism for the Net 2.0," *Mute: Culture and Politics after the Net,* September 21, 2006, http://www.metamute.org/en/Environmentalism-

for-Net-2.0. Together with Geert Lovink, Soenke Zehle set up the Web site incommunicado.net as a space to discuss and critique the global arrangements of the "information society." As part of this project, regular discussions of electronic waste and technology workers have appeared. See Geert Lovink and Soenke Zehle, eds., *Incommunicado Reader* (Amsterdam: Institute of Network Cultures, 2005); Matthias Feilhauer and Soenke Zehle, eds., "Ethics of Waste in the Information Society," special issue, *International Review of Information Ethics (IRIE)* 11 (October 2009).

37. Zehle, "Environmentalism for the Net 2.0."

38. There are an increasing number of projects that are operating within this area of digitally relevant environmentalism, which consider ways to address issues of environmental justice and green machines. For examples, see Shuzo Katsumoto, "Information and Communications Technology and the Environment: An Asian Perspective," *Journal of Industrial Ecology* 6, no. 2 (2003): 4–6; Jonathan Fildes, "Wireless Power System Shown Off," *BBC News*, July 23, 2009; Jonathan Fildes, "The Winds of Change for Africa," *BBC News*, July 23, 2009.

39. See Furtherfield, "Zero Dollar Laptop," http://www.furtherfield.org/zerodollarlaptop/; Access Space, http://www.access-space.org; Graham Harwood and Yokokoji Yoha, "Coal Fired Computers," *Discovery Museum* (Newcastle, United Kingdom: AV Festival, March 12–14, 2010).

40. Buchli, *Material Culture Reader*, 15. Jane Bennett similarly draws out the possibilities for thinking through new natures and new subjects that emerge through materializations. See Jane Bennett, "The Force of Things: Steps toward an Ecology of Matter," *Political Theory* 32, no. 3 (June 2004): 347–72.

41. Taussig, *Mimesis and Alterity*, 99.

42. Shanks, Platt, and Rathje, "Perfume of Garbage," 64.

43. Lynch, *Wasting Away*, 32.

44. Ibid., 41.

45. Benjamin, "Theses on the Philosophy of History," in *Illuminations*, 261–62.

Bibliography

Acland, Charles, ed. *Residual Media*. Minneapolis: University of Minnesota Press, 2007.

Active Disassembly. http://www.activedisassembly.com/.

Adkins, Lisa. "The New Economy, Property and Personhood." *Theory, Culture & Society* 22, no. 2 (2005): 111–30.

Adorno, Theodor W. "Natural History." In *Negative Dialectics*, translated by E. B. Ashton, 354–58. 1966. Reprint, London: Routledge, 1990.

Adorno, Theodor W. "A Portrait of Walter Benjamin." In *Prisms*, translated by Samuel Weber and Shierry Weber, 227–41. Cambridge, MA: MIT Press, 1967.

Agarwal, Ravi, and Kishore Wankhade. "Hi-Tech Heaps, Forsaken Lives: E-waste in Delhi." In *Challenging the Chip: Labor Rights and Environmental Justice in the Global Electronics Industry*, edited by Ted Smith, David A. Sonnenfeld, and David Naguib Pellow, 234–46. Philadelphia: Temple University Press, 2006.

Akrich, Madeline. "The De-Scription of Technical Objects." In *Shaping Technology/Building Society: Studies in Sociotechnical Change*, edited by Wiebe E. Bijker and John Law, 205–24. Cambridge, MA: MIT Press, 1992.

Alexander, Judd H. *In Defense of Garbage*. Westport, CT: Greenwood, 1993.

Alt, Franz. *Evaluation of Automatic Computing Machines*. Washington, DC: National Bureau of Standards, 1951.

Anderson, Ben, and John Wylie. "On Geography and Materiality." *Environment and Planning A*, 41, no. 2 (2009): 318–35.

Appadurai, Arjun. "Introduction: Commodities and the Politics of Value." In *The Social Life of Things: Commodities in Cultural Perspective*, edited by Arjun Appadurai, 3–63. Cambridge: Cambridge University Press, 1986.

Arendt, Hannah. "Introduction, Walter Benjamin: 1892–1940." In *Illuminations*, edited by Hanna Arendt, translated by Harry Zohn, 1–55. New York: Harcourt, Brace, Jovanovich, Schocken Books, 1969.

Babbage, Charles. *On the Economy of Machinery and Manufacturing*. London: Charles Knight, 1832.

Barney, Darin. *Prometheus Wired*. Vancouver: University of British Columbia, 2000.

Barthes, Roland. "Plastic." In *Mythologies*, translated by Annette Lavers, 97–99. New York: Farrar, Straus and Giroux, 1972.

Basel Action Network. *The Digital Dump: Exporting Re-use and Abuse to Africa*. Seattle: Basel Action Network, 2005.

Basel Action Network. http://www.ban.org.

Basel Action Network. "Detailed Critique of Problems with R2 Standard." November 2008. http://e-stewards.org/wp-content/uploads/2010/02/Detailed_R2_Analysis.pdf.

Basel Action Network and Silicon Valley Toxics Coalition. *Exporting Harm: The High-Tech Trashing of Asia*. Seattle and San Jose, 2002.

Bataille, Georges. *The Accursed Share: An Essay on General Economy*. Vol. 1, *Consumption*, translated by Robert Hurley. New York: Zone Books, 1991.

Bataille, Georges. "Dust." In *Encyclopedia Acephalica*, 42–43. London: Atlas Press, 1995.

Bataille, Georges. "Materialism." In *Visions of Excess: Selected Writings, 1927–1939*, translated by Alan Stoekl, 15–16. Minneapolis: University of Minnesota Press, 1985.

Baudrillard, Jean. *The Consumer Society: Myths and Structures*. Translated by Chris Turner. London: Sage, 1998.

Baudrillard, Jean. "The Remainder." In *Simulacra and Simulation*, translated by Sheila Faria Glaser, 143–48. Ann Arbor: University of Michigan Press, 1994.

Bauman, Zygmunt. *Liquid Modernity*. Cambridge: Polity Press, 2000.

Bauman, Zygmunt. *Wasted Lives: Modernity and Its Outcasts*. Cambridge: Polity Press, 2004.

Bell, Daniel. *The Coming of the Post-Industrial Society: A Venture in Social Forecasting*. New York: Basic Books, 1976.

Bell, Daniel. "The Social Framework of the Information Society." In *The Microelectronics Revolution: The Complete Guide to the New Technology and Its Impact on Society*, edited by Tom Forester, 500–549. Oxford: Basil Blackwell, 1980.

Bell, Gordon, and Jim Gemmell. *Total Recall: How the E-Memory Revolution Will Change Everything*. New York: Dutton, 2009.

Beniger, James R. *The Control Revolution: Technological and Economic Origins of the Information Society*. Cambridge: Harvard University Press, 1989.

Benjamin, Walter. *The Arcades Project*. Translated by Howard Eiland and Kevin McLaughlin. Cambridge: Belknap Press of Harvard University Press, 1999.

Benjamin, Walter. "Excavation and Memory." In *Selected Writings*, vol. 2, *1931–1934*, edited by Michael W. Jennings, Howard Eiland, and Gary Smith, translated by Rodney Livingstone et al., 576. Cambridge: Belknap Press of Harvard University Press, 1999.

Benjamin, Walter. "One-Way Street." In *Reflections*, edited by Peter Demetz, translated by Edmund Jephcott, 61–94. New York: Harcourt, Brace, Jovanovich, Schocken Books, 1986.

Benjamin, Walter. *The Origin of German Tragic Drama*. Translated by John Osborne. London: NLB, 1977.

Benjamin, Walter. "The Paris of the Second Empire in Baudelaire." In *Selected Writings*, vol. 4, *1938–1940*, edited by Howard Eiland and Michael W. Jennings, translated by Edmund Jephcott et al., 3–92. Cambridge: Belknap Press of Harvard University Press, 2002.

Benjamin, Walter. "The Storyteller: Reflections on the Works of Nikolai Leskov." In *Illuminations*, edited by Hannah Arendt, translated by Harry Zohn, 83–109. New York: Harcourt, Brace, Jovanovich, Schocken Books, 1969.

Benjamin, Walter. "Surrealism: The Last Snapshot of the European Intelligentsia." In *Reflections*, edited by Peter Demetz, translated by Edmund Jephcott, 177–92. New York: Harcourt, Brace, Jovanovich, Schocken Books, 1986.

Benjamin, Walter. "Theses on the Philosophy of History." In *Illuminations*, edited by Hannah Arendt, translated by Harry Zohn, 253–64. New York: Harcourt, Brace, Jovanovich, Schocken Books, 1969.

Benjamin, Walter. "The Work of Art in the Age of Mechanical Reproduction." In *Illuminations*, edited by Hannah Arendt, translated by Harry Zohn, 217–51. New York: Harcourt, Brace, Jovanovich, Schocken Books, 1969.

Bennett, Jane. "The Force of Things: Steps toward an Ecology of Matter." *Political Theory* 32, no. 3 (June 2004): 347–72.

Bensaude-Vincent, Bernadette, and Isabelle Stengers. *A History of Chemistry*. Translated by Deborah van Dam. Cambridge: Harvard University Press, 1996.

Berkeley, Edmund. *Giant Brains, or Machines That Think*. New York: John Wiley and Sons, 1949.

Berlin, Leslie. *The Man Behind the Microchip: Robert Noyce and the Invention of Silicon Valley*. Oxford: Oxford University Press, 2005.

Bijker, Wiebe E., and John Law, eds. *Shaping Technology/Building Society: Studies in Sociotechnical Change*. Cambridge, MA: MIT Press, 1992.

Bolter, J. David. *Turing's Man: Western Culture in the Computer Age*. Chapel Hill: University of North Carolina Press, 1984.

Borgmann, Albert. *Holding on to Reality: The Nature of Information at the Turn of the Millennium*. Chicago: University of Chicago Press, 1999.

Boyle, James. "A Politics of Intellectual Property: Environmentalism for the Net?" 1997. http://www.law.duke.edu/boylesite/Intprop.htm.

Brand, Stewart. *The Clock of the Long Now: Time and Responsibility*. New York: Basic Books, 1999.

Brown, Bill. "Thing Theory." *Critical Inquiry* 28, no. 1, Things (Autumn 2001): 1–22.

Brucker-Cohen, Jonah. "Scrapyard Challenge Workshops." http://infamia1.infamia.com/coin-operated.com.

Buchli, Victor. Introduction to *The Material Culture Reader*, edited by Victor Buchli, 1–22. Oxford: Berg, 2002.

Buck-Morss, Susan. *The Dialectics of Seeing: Walter Benjamin and the Arcades Project*. Cambridge, MA: MIT Press, 1989.

Bush, Vannevar. "As We May Think." *Atlantic Monthly*, July 1945, 101–8.

Butler, Judith. *Bodies That Matter: On the Discursive Limits of "Sex."* New York: Routledge, 1993.

Callon, Michel. "What Does It Mean to Say That Economics Is Performative?" In *Do Economists Make Markets? On the Performativity of Economics*, edited by Donald MacKenzie, Fabian Muniesa, and Lucia Siu, 311–57. Princeton: Princeton University Press, 2007.

Callon, Michel, and John Law. "On Qualculation, Agency and Otherness." *Environment and Planning D: Society and Space* 23, no. 5 (2005), 717–33.

Calvino, Italo. *Invisible Cities*. Translated by William Weaver. New York: Harcourt, Brace, Jovanovich, 1974.

Campbell-Kelly, Martin. *From Airline Reservations to Sonic the Hedgehog: A History of the Software Industry*. Cambridge, MA: MIT Press, 2004.

Carrier, James G., and Daniel Miller. *Virtualism: A New Political Economy*. Oxford: Berg, 1998.

Castells, Manuel. *The Rise of the Network Society*. Oxford: Blackwell, 2000.

CERCLA. "Priority List of Hazardous Substances." http://www.atsdr.cdc.gov/cercla/index.asp.

Ceruzzi, Paul. *A History of Modern Computing*. Cambridge, MA: MIT Press, 2003.
Chun, Wendy Hui Kyong, and Thomas Keenan, eds. *New Media, Old Media*. New York: Routledge, 2006.
Clapp, Jennifer. *Toxic Exports*. Ithaca: Cornell University Press, 2001.
Clark, Gordon L., and Nigel Thrift. "Performing Finance: The Industry, the Media and Its Image." *Review of International Political Economy* 11, no. 2 (May 2004): 289–310.
Clarke, Arthur C. *Profiles of the Future: An Inquiry into the Limits of the Possible*. London: Indigo, 1999.
Colloredo-Mansfeld, Rudi. "Consuming Andean Televisions." *Journal of Material Culture* 8, no. 3 (2003): 273–84.
Colloredo-Mansfeld, Rudi. "Introduction: Matter Unbound." *Journal of Material Culture* 8, no. 3 (2003): 245–54.
Core 77. "Greener Gadgets Design Competition." http://www.core77.com/competitions/GreenerGadgets.
Corzo, Miguel Angel, ed. *Mortality/Immortality? The Legacy of 20th-Century Art*. Los Angeles: Getty Conservation Institute, 1999.
Couldry, Nick, and Anna McCarthy, eds. *MediaSpace: Place, Scale and Culture in a Media Age*. London: Routledge, 2004.
Currimbhoy, N. "New NASDAQ MarketSite Design Inspired by Computer Chip (New-York-City Headquarters Resembles the Inside of a Computer)." *Architectural Record* 186, no. 5 (May 1998): 262.
Darwin, Charles. *On the Origin of the Species*. London: John Murray, 1859.
Dell. "Corporate Responsibility Report." 2009. http://www.dell.com/sustainabilityreport.
Derrida, Jacques. *Archive Fever: A Freudian Impression*, translated by Eric Prenowitz. Chicago: University of Chicago Press, 1996.
Dick, Philip K. *In Milton Lumky Territory*. 1985. Reprint, London: Paladin Books, 1987.
Diebold, John. *Automation: The Advent of the Automatic Factory*. New York: D. Van Nostrand., 1952.
Diebold, John. "Factories without Men: New Industrial Revolution." *Nation*, September 19, 1953.
"Directive 2002/96/EC of the European Parliament and of the Council of 27 January 2003 on Waste Electrical and Electronic Equipment (WEEE)." *Official Journal of the European Union*, February 13, 2003, L37/24–L37/38.
"Directive 2002/95/EC of the European Parliament and of the Council of 27 January 2003 on the Restriction of the Use of Certain Hazardous Substances in Electrical and Electronic Equipment." *Official Journal of the European Union*, February 13, 2003, L37/19–L37/23.
DiSalvo, Carl, Kirsten Boehner, Nicholas A. Knouf, and Phoebe Sengers. "Nourishing the Ground for Sustainable HCI: Considerations from Ecologically Engaged Art." *Proceedings of the CHI Conference on Human Computing Systems*, April 4–9, 2009, Boston.
Dooren, Noel van. "Never Again Will the Heap Lie in Peace." In *Tales of the Tip*, edited by Chris Driessen and Hiedi van Mierlo, 100–105. Amsterdam: Fundament Foundation, 1999.

Douglas, Mary. *Purity and Danger: An Analysis of Concepts of Pollution and Taboo*. 1966. Reprint, London: Routledge, 1995.
Dunne, Anthony. *Hertzian Tales: Electronic Products, Aesthetic Experience and Critical Design*. Cambridge, MA: MIT Press, 2005.
Dupuy, Jean-Pierre. "Myths of the Informational Society." In *The Myths of Information: Technology and Postindustrial Culture*, edited by Kathleen Woodward, 3–17. Madison, WI: Coda Press, 1980.
Dyer-Witheford, Nick. *Cyber-Marx: Cycles and Circuits of Struggle in High-Technology Capitalism*. Urbana: University of Illinois Press, 1999.
Eames, Charles, and Ray Eames. *A Computer Perspective: Background to the Computer Age*. Cambridge, MA: Harvard University Press, 1990.
Electronic Product Environmental Assessment Tool. http://www.epeat.net.
"Electronic Waste Guide." http://www.ewaste.ch/.
Engler, Mira. *Designing America's Waste Landscapes*. Baltimore: Johns Hopkins University Press, 2004.
Environmental Protection Agency. "Fairchild Semiconductor Case Study." http://epa.gov/superfund/programs/recycle/live/casestudy_fairchild.html.
Environmental Protection Agency. "National Priorities List." http://www.epa.gov/superfund/sites/npl.
Environmental Protection Agency. "National Priorities List Sites in California" (Santa Clara County). http://yosemite.epa.gov/r9/sfund/r9sfdocw.nsf/WSOState!OpenView&Start=1&Count=1000&Expand=2.29#2.29.
Environmental Protection Agency. "Responsible Recycling ('R2') Practices for Use in Accredited Certification Programs for Electronics Recyclers." October 30, 2008. http://www.epa.gov/waste/conserve/materials/ecycling/r2practices.htm.
Environmental Protection Agency. Superfund Sites. http://www.epa.gov/region09/superfund/superfundsites.html.
Environmental Protection Agency. Toxic Release Inventory Program. http://www.epa.gov/tri.
Environmental Protection Agency. "Waste Wise Update: Electronics Reuse and Recycling." EPA 530-N-00-007. October 2000. http://www.epa.gov/epawaste/partnerships/wastewise/pubs/wwupda14.pdf.
Ernst, Wolfgang. "Agencies of Cultural Feedback: The Infrastructure of Memory." In *Waste-Site Stories: The Recycling of Memory*, edited by Brian Neville and Johanne Villeneuve, 107–20. Albany: State University of New York Press, 2002.
Ernst, Wolfgang. "Archi(ve)textures of Museology." In *Museums and Memory*, edited by Susan A. Crane, 17–34. Stanford: Stanford University Press, 2000.
E-Stewards Initiative. http://www.e-stewards.org/.
Featherstone, Mike. "Archiving Cultures." *British Journal of Sociology* 51, no. 1 (January–March 2000): 161–84.
Feilhauer, Matthias, and Soenke Zehle, eds. "Ethics of Waste in the Information Society." Special issue, *International Review of Information* Ethics (IRIE) 11 (October 2009).
Ferrell, Jeff. *Empire of Scrounge: Inside the Urban Underground of Dumpster Diving, Trash Picking, and Street Scavenging*. New York: New York University Press, 2006.
Finn, Christine A. *Artifacts: An Archaeologist's Year in Silicon Valley*. Cambridge, MA: MIT Press, 2001.

Fishbein, Bette K. *Waste in a Wireless World: The Challenge of Cell Phones*. New York: Inform, 2002.

Fishbein, Bette K., et al. *Extended Producer Responsibility: A Materials Policy for the 21st Century*. New York: Inform, 2000.

Fisher, Melissa, and Greg Downey. "Introduction: The Anthropology of Capital and the Frontiers of Ethnography." In *Frontiers of Capital: Ethnographic Reflections on the New Economy*, edited by Melissa Fisher and Greg Downey, 1–30. Durham: Duke University Press, 2006.

Fladager, Vernon L. *The Selling Power of Packaging*. New York: McGraw-Hill, 1956.

Forty, Adrian. *Objects of Desire: Design and Society, 1750–1980*. London: Thames and Hudson, 1986.

Forty, Adrian, and Susanne Küchler, eds. *The Art of Forgetting*. Oxford: Berg, 1999.

Foucault, Michel. *The Order of Things: An Archaeology of the Human Sciences*. 1970. Reprint, London: Routledge, 1994.

Franklin, Sarah, Celia Lury, and Jackie Stacey. *Global Nature, Global Culture*. London: Sage, 2000.

Frey, Sibylle D., David J. Harrison, and Eric H. Billett. "Ecological Footprint Analysis Applied to Mobile Phones." *Journal of Industrial Ecology* 10, nos. 1–2 (2006): 199–216.

Froth, Marcus, Eric Paulos, Christine Satchell, and Paul Dourish. "Pervasive Computing and Environmental Sustainability: Two Conference Workshops." *IEEE CS* 8, no. 1 (January–March 2009): 78–81.

Frow, John. "Invidious Distinction: Waste, Difference, and Classy Stuff." In *Culture and Waste: The Creation and Destruction of Value*, edited by Gay Hawkins and Stephen Muecke, 25–38. Lanham, MD: Rowman and Littlefield, 2003.

Frow, John. "A Pebble, a Camera, a Man Who Turns into a Telegraph Pole." *Critical Inquiry* 28, no. 1, Things (Autumn 2001): 270–85.

Fuad-Luke, Alastair. *The Eco-Design Handbook*. London: Thames and Hudson, 2005.

Fuller, Matthew. *Media Ecologies: Materialist Energies in Art and Technoculture*. Cambridge, MA: MIT Press, 2005.

Furtherfield. "Zero Dollar Laptop." http://www.furtherfield.org/zerodollarlaptop/.

Gabrys, Jennifer. "Machines Fall Apart: Failure in Art and Technology." *Leonardo Electronic Almanac* 13, no. 4 (April 2005). http://www.leoalmanac.org/journal/Vol_13/lea_v13_n04.txt.

Gabrys, Jennifer. "Paper Mountains, Disposable Cities." In *Surface Tension Supplement* 1. Edited by Brandon Labelle and Ken Ehrlich, 130–39. Los Angeles: Errant Bodies Press, 2006.

Gabrys, Jennifer. "Sink: The Dirt of Systems." *Environment and Planning D: Society and Space* 27, no. 4 (2009): 666–81.

Gadrey, Jean. *New Economy, New Myth*. 2001. London: Routledge, 2003.

Gandy, Matthew. *Recycling and the Politics of Urban Waste*. New York: St. Martin's Press, 1994.

Gaonkar, Dilip Parameshwar, and Elizabeth A. Povinelli. "Technologies of Public Forms: Circulation, Transfiguration, Recognition." *Public Culture* 15, no. 3 (2003): 385–97.

Gaulon, Benjamin. "Recyclism." http://www.recyclism.com/.

Geiser, Kenneth. "The Chips Are Falling: Health Hazards in the Microelectronics Industry." *Science for the People* 17, no. 8 (1985).

Geiser, Kenneth. *Materials Matter: Towards a Sustainable Materials Policy*. Cambridge, MA: MIT Press, 2001.

Gemmell, Jim, Gordon Bell, and Roger Lueder. "MyLifeBits: A Personal Database for Everything." Microsoft Research Technical Report, MSR-TR-2006-23. San Francisco: Microsoft Bay Area Research Center, 2006. http://research.microsoft.com/apps/pubs/default.aspx?id=64157.

Gitelman, Lisa. *Always Already New: Media, History, and the Data of Culture*. Cambridge, MA: MIT Press, 2006.

Gitlin, Todd. *Media Unlimited: How the Torrent of Images and Sounds Overwhelms Our Lives*. New York: Henry Holt, 2003.

Gould, Stephen Jay. *Wonderful Life: The Burgess Shale and the Nature of History*. 1990. Reprint, London: Vintage Books, 2000.

Graham, Stephen, and Nigel Thrift. "Out of Order: Understanding Repair and Maintenance." *Theory, Culture & Society* 24, no. 3 (2007): 1–25.

Grattan, David, and R. Scott Williams. "From '91' to '42': Questions of Conservation for Modern Materials." In *Mortality/Immortality? The Legacy of 20th-Century Art*, edited by Miguel Angel Corzo, 67–74. Los Angeles: Getty Conservation Institute, 1999.

Green Electronics Council with the National Center for Electronics Recycling and Resource Recycling. "Closing the Loop: Electronics Design to Enhance Reuse/Recycling Value." January 2009. http://www.greenelectronicscouncil.org/documents/0000/0007/Design_for_End_of_Life_Final_Report_090208.pdf.

Greenpeace. "Greenpeace Pulls Plug on Dirty Electronics Companies." May 23, 2005. http://www.greenpeace.org/international/en/press/releases/greenpeace-pulls-plug-on-dirty.

Greenpeace. "Guide to Greener Electronics." December 2008. http://www.greenpeace.org/rankingguide.

Greenpeace. "Recycling of Electronic Wastes in China & India: Workplace & Environmental Contamination." August 2005. http://www.greenpeace.to/publications/electronic_waste_recycling_appendix.pdf.

Grossman, Elizabeth. *High Tech Trash: Digital Devices, Hidden Toxics, and Human Health*. Washington, DC: Island Press, 2006.

Grosz, Elizabeth. *Architecture from the Outside*. Cambridge, MA: MIT Press, 2001.

Guattari, Felix. *The Three Ecologies*. Translated by Ian Pindar and Paul Sutton. London: Athlone Press, 2000.

Gumbrecht, Hans Ulrich, and Michael Marrinan, eds. *Mapping Benjamin: The Work of Art in the Digital Age*. Stanford: Stanford University Press, 2003.

Gumbrecht, Hans Ulrich, and K. Ludwig Pfeiffer, eds. *Materialities of Communication*, translated by William Whobrey. Stanford: Stanford University Press, 1994.

Haggerty, Patrick. "Integrated Electronics: A Perspective." In *Management Philosophies and Practices of Texas Instruments*. Dallas: Texas Instruments, 1965. Reprinted in Frederick Seitz, and Norman G. Einspruch. *Electronic Genie: The Tangled History of Silicon*. Urbana: University of Illinois Press, 1998.

Hally, Mike. *Electronic Brains: Stories from the Dawn of the Computer Age*. London: Granta Books, 2005.

Hammond, Allen, et al. *Resource Flows: The Material Basis of Industrial Economies*. Washington, DC: World Resources Institute, 1997.

Hansen, Karen Tranberg. *Salaula: The World of Secondhand Clothing and Zambia.* Chicago: University of Chicago Press, 2000.

Hansen, Mark. *Embodying Technesis: Technology beyond Writing.* Ann Arbor: University of Michigan Press, 2000.

Hanssen, Beatrice. *Walter Benjamin's Other History: Of Stones, Animals, Human Beings, and Angels.* Berkeley: University of California Press, 1998.

Haraway. Donna. "A Cyborg Manifesto: Science, Technology, and Socialist-Feminism in the Late Twentieth Century." In *Simians, Cyborgs, and Women: The Reinvention of Nature,* 149–81. New York: Routledge, 1991.

Haraway, Donna. "Cyborgs, Coyotes, and Dogs: A Kinship of Feminist Figurations" and "There are Always More Things Going on than You Thought! Methodologies as Thinking Technologies: An Interview with Donna Haraway," conducted in two parts by Nina Lykke, Randi Markussen, and Finn Olesen. In *The Haraway Reader,* 321–42. New York: Routledge, 2004.

Haraway, Donna. "Cyborgs to Companion Species: Reconfiguring Kinship in Technoscience." In *The Haraway Reader,* 295–320. New York: Routledge, 2004.

Haraway, Donna. *How Like a Leaf: An Interview with Thyrza Nichols Goodeve.* New York: Routledge, 2000.

Haraway, Donna. *Modest_Witness@Second_Millennium. FemaleMan©_Meets_Onco Mouse™.* New York: Routledge, 1997.

Haraway, Donna. "The Promises of Monsters: A Regenerative Politics for Inappropriate/d Others." In *The Haraway Reader,* 63–124. New York: Routledge, 2004.

Haraway, Donna. *Simians, Cyborgs, and Women: The Reinvention of Nature.* London: Free Association Books, 1991.

Haraway, Donna. "Situated Knowledges: The Science Question in Feminism and the Privilege of Partial Perspective." *Feminist Studies* 14, no. 3 (1988): 575–99.

Harpold, Terry, and Kavita Philip. "Of Bugs and Rats: Cyber-Cleanliness, Cyber-Squalor, and the Fantasy-Spaces of Informational Globalization." *Postmodern Culture* 11, no. 1 (2000). http://muse.jhu.edu/journals/pmc/v011/11.1harpold.html.

Harvey, David. *Spaces of Capital: Towards a Critical Geography.* London: Routledge, 2001.

Harwood, Graham, and Matsuko Yokokoji. "Coal Fired Computers." *Discovery Museum.* Newcastle, United Kingdom: AV Festival, March 12–14, 2010.

Hawkins, Gay. *The Ethics of Waste: How We Relate to Rubbish.* Lanham, MD: Rowman and Littlefield, 2005.

Hawkins, Gay. "Plastic Bags: Living with Rubbish." *International Journal of Cultural Studies* 4, no. 1 (2001): 5–23.

Hawkins, Gay, and Stephen Muecke, eds. *Culture and Waste: The Creation and Destruction of Value.* Lanham, MD: Rowman and Littlefield, 2003.

Hayles, N. Katherine. *Chaos Bound: Orderly Disorder in Contemporary Literature and Science.* Ithaca: Cornell University Press, 1990.

Hayles, N. Katherine. *How We Became Posthuman: Virtual Bodies in Cybernetics, Literature, and Informatics.* Chicago: University of Chicago Press, 1999.

Hayles, N. Katherine. *My Mother Was a Computer.* Chicago: University of Chicago Press, 2005.

Henwood, Doug. *After the New Economy: The Binge and the Hangover That Won't Go Away.* New York: New Press, 2005.

Hetherington, Kevin. "Secondhandedness: Consumption, Disposal, and Absent Presence." *Environment and Planning D: Society and Space* 22, no. 1 (2004): 157–73.

Hetherington, Kevin, and John Law. "After Networks." *Environment and Planning D: Society and Space* 18, no. 2 (2000): 127–32.

Hinte, Ed van. *Eternally Yours: Visions on Product Endurance*. Rotterdam: 010 Publishers, 2004.

Horn, Eva. "There Are No Media." *Grey Room* 29 (Winter 2008): 6–13.

"How Much Information." http://hmi.ucsd.edu/howmuchinfo.php.

"How Stuff Is Made." http://www.howstuffismade.org.

Huh, Jina, et al. "Workshop on Examining Appropriation, Re-use, and Maintenance for Sustainability." http://jinah.people.si.umich.edu//chi2010/reuse.html.

IDC. "The Diverse and Exploding Digital Universe." An IDC Whitepaper, sponsored by EMC. Framingham, MA: IDC, 2008.

Institute of Scrap Recycling Industries. http://www.isri.org.

Intel. "4004: Intel's First Microprocessor." http://intelpr.feedroom.com/.

Intel. "From Sand to Circuits: How Intel Makes Integrated Circuit Chips." 2008. ftp://download.intel.com/museum/sand_to_circuits.pdf.

Intel. "Invention, Innovation, Investment." http://intelpr.feedroom.com/.

International Association of Electronics Recyclers. *IAER Electronics Recycling Industry Report*. Albany: International Association of Electronics Recyclers, 2003, revised 2004.

Internet Archive. http://www.archive.org.

Jeremijenko, Natalie, and Proboscis. "Feral Robotic Dogs." http://www.nyu.edu/projects/xdesign/feralrobots/.

Kabakov, Ilya. "The Man Who Never Threw Anything Away." In *The Archive*, edited by Charles Merewether, 32–37. London: Whitechapel Gallery; Cambridge, MA: MIT Press, 2006.

Katsumoto, Shuzo. "Information and Communications Technology and the Environment: An Asian Perspective." *Journal of Industrial Ecology* 6, no. 2 (2003): 4–6

Kelly, Kevin. *Out of Control: The New Biology of Machines, Social Systems, and the Economic World*. New York: Basic Books, 1994.

Kittler, Friedrich A. *Discourse Networks, 1800/1900*. Translated by Michael Metteer and Chris Cullens. Stanford: Stanford University Press, 1990.

Kittler, Friedrich A. "The Exhibition as Emulator." Translated by James Boekbinder. Text commissioned for the *InfoArcadia* exhibition, 2000. http://www.mediamatic.net/article-8740-en.html.

Kittler, Friedrich A. "Museums on the Digital Frontier." In *The End(s) of the Museum*, edited by Alexander Garcia Düttmann et al., 67–80. Barcelona: Fundació Antoni Tàpies, 1996.

Kittler, Friedrich A. "There Is No Software." *CTHEORY* (October 18, 1995). http://www.ctheory.net/articles.aspx?id=74.

Knorr-Cetina, Karin. "From Pipes to Scopes: The Flow Architecture of Financial Markets." In *The Technological Economy*, edited by Andrew Barry and Don Slater, 122–41. London: Routledge, 2005.

Knorr-Cetina, Karin, and Urs Bruegger. "Global Microstructures: The Virtual Societies of Financial Markets." *American Journal of Sociology* 107, no. 4 (2002): 905–50.

Köhler, Andreas, and Lorenz Erdmann. "Expected Environmental Impacts of Pervasive Computing." *Human and Ecological Risk Assessment* 10, no. 5 (October 2004): 831–52.

Koolhaas, Rem. "Junkspace."*October* 100 (Spring 2002): 175–90.
Koolhaas, Rem. "Wasteland. Dump Space: Freedom from Order." *Wired* 11, no. 6 (June 2003).
Krauss, Rosalind. *A Voyage on the North Sea: Art in the Age of the Post-Medium Condition.* London: Thames and Hudson, 1999.
Krishna, Gopal. "E-Waste: Computers and Toxicity in India." *Sarai 3: Shaping Technologies* (2003): 12–13.
Kuehr, Ruediger, German T. Velasquez, and Eric Williams. "Computers and the Environment: An Introduction to Understanding and Managing Their Impacts." In *Computers and the Environment: Understanding and Managing their Impacts*, edited by Ruediger Kuehr and Eric Williams, 1–16. Dordrecht: Kluwer Academic, 2003.
Kuehr, Ruediger, and Eric Williams, eds. *Computers and the Environment: Understanding and Managing Their Impacts.* Dordrecht: Kluwer Academic, 2003.
Kuhn, Thomas S. *The Structure of Scientific Revolutions.* 1962. Reprint, Chicago: University of Chicago Press, 1996.
Lash, Scott. *Critique of Information.* London: Sage, 2002.
Lash, Scott, and Celia Lury. *Global Culture Industry.* Cambridge: Polity Press, 2007.
Latham, Robert, and Saskia Sassen, eds. *Digital Formations: IT and New Architectures in the Global Realm.* Princeton: Princeton University Press, 2005.
Latour, Bruno. *Aramis, or the Love of Technology.* Translated by Catherine Porter. Cambridge: Harvard University Press, 1996.
Latour, Bruno. *Pandora's Hope: Essays on the Reality of Science Studies.* Cambridge: Harvard University Press.
Latour, Bruno. *Reassembling the Social: An Introduction to Actor-Network-Theory.* Oxford University Press, 2005.
Latour, Bruno. "Technology Is Society Made Durable." In *A Sociology of Monsters: Essays on Power, Technology and Domination*, edited by John Law, 103–31. London: Routledge, 1991.
Latour, Bruno. *We Have Never Been Modern.* Translated by Catherine Porter. Cambridge: Harvard University Press, 1993.
Law, John, and Annemarie Mol. "Notes on Materiality and Sociality." *Sociological Review* 43 (1995): 274–94.
Lee, Benjamin, and Edward LiPuma. "Cultures of Circulation: The Imaginations of Modernity." *Public Culture* 14, no. 1 (2002): 191–213.
Lécuyer, Christophe. *Making Silicon Valley: Innovation and the Growth of High Tech, 1930–1970.* Cambridge, MA: MIT Press, 2005.
Lécuyer, Christophe, and David C. Brock. "The Materiality of Microelectronics." *History and Technology* 22, no. 3 (September 2006): 301–25.
Leslie, Esther. *Synthetic Worlds: Nature, Art and the Chemical Industry.* London: Reaktion Books, 2005.
Levinson, Marc. *The Box: How the Shipping Container Made the World Smaller and the World Economy Bigger.* Princeton: Princeton University Press, 2006.
Levinson, Paul. *The Soft Edge: A Natural History and Future of the Information Revolution.* New York: Routledge, 1997.
The Long Now Foundation. http://www.longnow.org.
Loon, Joost van, and Ida Sabelis. "Recycling Time. The Temporal Complexity of Waste Management." *Time & Society* 6, no. 2 (June 1997): 287–306.
Lovink, Geert. "Archive Rumblings: Interview with German Media Archaeologist

Wolfgang Ernst." Nettime, February 25, 2003. http://www.nettime.org/Lists-Archives/nettime-l-0302/msg00132.html.

Lovink, Geert, and Soenke Zehle, eds. *Incommunicado Reader*. Amsterdam: Institute of Network Cultures, 2005.

Lucas, Gavin. "Disposability and Dispossession in the Twentieth Century." *Journal of Material Culture* 7, no. 1 (2002): 5–22.

Lukács, Georg. *Theory of the Novel*. Cambridge, MA: MIT Press, 1974.

Lupton, Ellen, and J. Abbott Miller. *The Bathroom, the Kitchen and the Aesthetics of Waste: A Process of Elimination*. Cambridge, MA: MIT List Visual Arts Center, 1992.

Lyman, Peter, and Hal R. Varian. "How Much Information." 2003. http://www2.sims.berkeley.edu/research/projects/how-much-info-2003/index.htm.

Lynch, Kevin. *Wasting Away*. San Francisco: Sierra Club Books, 1990.

Lyotard, Jean-François. *Driftworks*. Edited by Roger McKeon. New York: Semiotext(e), 1984.

Lyotard, Jean-François. *The Inhuman: Reflections on Time*. Translated by Geoffrey Bennington and Rachel Bowlby. Cambridge: Polity Press, 1991.

Lyotard, Jean-François. *The Postmodern Condition: A Report on Knowledge*. Translated by Brian Massumi. Minneapolis: University of Minnesota Press, 1985.

Machlup, Fritz. *Knowledge and Knowledge Production (Knowledge: Its Creation, Distribution, and Economic Significance)*. Vol. 1. Princeton: Princeton University Press, 1980.

MacKay, Donald M. *Information, Mechanism, Meaning*. Cambridge, MA: MIT Press, 1969.

Mackenzie, Adrian. "The Performativity of Code: Softwares and Cultures of Circulation." *Theory, Culture, and Society* 22, no. 1 (2005): 71–92.

Mackenzie, Adrian. "These Things Called Systems: Collective Imaginings and Infrastructural Software." *Social Studies of Science* 33, no. 3 (June 2003): 365–87.

MacKenzie, Donald. *An Engine, Not a Camera: How Financial Models Shape Markets*. Cambridge, MA: MIT Press, 2006.

MacKenzie, Donald. "Is Economics Performative? Option Theory and the Construction of Derivative Markets." In *Do Economists Make Markets? On the Performativity of Economics*, edited by Donald MacKenzie, Fabian Muniesa, and Lucia Siu, 54–86. Princeton: Princeton University Press, 2007.

MacKenzie, Donald. *Knowing Machines: Essays on Technical Change*. Cambridge, MA: MIT Press, 1998.

MacKenzie, Donald, Fabian Muniesa, and Lucia Siu, eds. *Do Economists Make Markets? On the Performativity of Economics*. Princeton: Princeton University Press, 2007.

Manzini, Ezio. *The Material of Invention*. London: Design Council, 1986.

Marker, Chris. *Sans Soleil*. 1982.

Marx, Karl. *Capital: A Critique of Political Economy* vol. 1. Translated by Ben Fowkes. London: Penguin Books, 1990.

Marx, Karl. *Grundrisse: Foundations of the Critique of Political Economy* (rough draft). Translated by Martin Nicolaus. London: Pelican, 1973.

Marx, Ursula, et al., eds. "Ragpicking: *The Arcades Project*." In *Walter Benjamin's Archive: Images, Texts, Signs*, translated by Esther Leslie, 251–65. London: Verso, 2007.

Massumi, Brian. *Parables for the Virtual: Movement, Affect, Sensation*. Durham: Duke University Press, 2002.

Matthews, Emily, et al. *The Weight of Nations: Material Outflows from Industrial Economies.* Washington, DC: World Resources Institute, 2000.

Matthews, H. Scott, and Deanna Matthews. "Information Technology Products and the Environment." In *Computers and the Environment*, edited by Ruediger Kuehr and Eric Williams, 17–40. Dordrecht: Kluwer Academic, 2003.

Mayers, C. Kieren, Chris M. France, and Sarah J. Cowell. "Extended Producer Responsibility for Waste Electronics." *Journal of Industrial Ecology* 9, no. 3 (2005): 169–89.

Mazurek, Jan. *Making Microchips: Policy, Globalization, and Economic Restructuring in the Semiconductor Industry*. Cambridge, MA: MIT Press, 1999.

McCarthy, James E. "Recycling Computers and Electronic Equipment: Legislative and Regulatory Approaches for 'E-Waste.'" Report for Congress, CRS-1, July 19, 2002.

McDonough, William, and Michael Braungart. *Cradle to Cradle: Remaking the Way We Make Things.* New York: North Point Press, 2002.

McElvenny, Damien M., et al. "Cancer among Current and Former Workers at National Semiconductor (UK) Ltd., Greenock." Norwich, UK: Health and Safety Executive Books, 2001.

McLuhan, Marshall. "Automation: Learning a Living." In *Understanding Media*, 346–59. Cambridge, MA: MIT Press, 1994.

McLuhan, Marshall. *Understanding Media*. Cambridge, MA: MIT Press, 1994.

McLuhan, Marshall, and Quentin Fiore. *The Medium Is the Massage: An Inventory of Effects*. New York: Bantam Books, 1967.

McPherson, Alexandra, Beverley Thorpe, and Ann Blake. "Brominated Flame Retardants in Dust on Computers: The Case for Safer Chemicals and Better Computer Design." June 2004. http://www.electronicstakeback.com.

Meikle, Jeffrey L. *American Plastic: A Cultural History*. New Brunswick, NJ: Rutgers University Press, 1995.

Michael, Mike. *Reconnecting Culture, Technology and Nature: From Society to Heterogeneity.* London: Routledge, 2000.

Miller, Daniel, ed. *Materiality.* Durham: Duke University Press, 2005.

Mitchell, Timothy. "The Character of Calculability." In *Rule of Experts: Egypt, Techno-Politics, Modernity*, 80–122. Berkeley: University of California Press, 2002.

Mitchell, Timothy. "Rethinking Economy." *Geoforum* 39, no. 3 (2008): 1116–21.

Mitchell, William J. *City of Bits: Space, Place, and the Infobahn.* Cambridge, MA: MIT Press, 1995.

Miyazaki, Hirokazu. "The Materiality of Finance Theory." In *Materiality*, edited by Daniel Miller, 165–81. Durham: Duke University Press, 2005.

"Molded Plastic Containers." *Modern Packaging Journal* 31, no. 1 (September 1957): 120–23, 240–41.

Molotch, Harvey. *Where Stuff Comes From: How Toasters, Toilets, Cars, Computers, and Many Other Things Come to Be as They Are*. New York: Routledge, 2003.

Moore, Gordon E. "Cramming More Components onto Integrated Circuits." *Electronics* 38, no. 8 (April 19, 1965): 114–17.

Moore, Gordon E. Interview, March 3, 1995. In *Silicon Genesis: An Oral History of Semiconductor Technology.* Stanford and the Silicon Valley Project. http://silicongenesis.stanford.edu/complete_listing.html.

Moore, Gordon E. "No Exponential Is Forever . . . but We Can Delay Forever." *Solid State Circuits Conference Proceedings* 1 (2003): 20–23.

Moser, Walter. "The Acculturation of Waste." In *Waste-Site Stories: The Recycling of Memory*, edited by Brian Neville and Johanne Villeneuve, 85–106. Albany: State University of New York Press, 2002.

Muniesa, Fabian. "Assemblage of a Market Mechanism." *Journal of the Center for Information Studies* 5, no. 3 (2004): 11–19.

Munster, Anna. *Materializing New Media: Embodiment in Media Aesthetics*. Hanover, NH: Dartmouth College Press, 2006.

Murphy, Michelle. *Sick Building Syndrome and the Problem of Uncertainty: Environmental Politics, Technoscience, and Women Workers*. Durham: Duke University Press, 2006.

Murray, Robin. *Zero Waste*. London: Greenpeace Environmental Trust, 2002.

"Museum of E-Failure." http://www.disobey.com/ghostsites/.

Nas, Peter J. M., and Rivke Jaffe. "Informal Waste Management: Shifting the Focus from Problem to Potential." *Environment, Development and Sustainability* 6 (2004): 337–53.

NASDAQ. "Built for Business: Annual Report, 2004." http://ir.nasdaq.com/annuals.cfm.

NASDAQ. "MarketSite Fact Sheet." http://www.nasdaq.com/reference/marketsite_facts.stm.

NASDAQ. "Performance Report." http://www.nasdaq.com/newsroom/stats/Performance_Report.stm. Accessed March 4, 2008.

NASDAQ. "2007 Annual Report." http://ir.nasdaq.com/annuals.cfm.

National Oceanic and Atmospheric Administration. "Maritime Shipping Makes Hefty Contribution to Harmful Air Pollution." February 26, 2009. http://www.noaanews.noaa.gov/stories2009/20090226_shipping.html.

National Safety Council. *Electronic Product Recovery and Recycling Baseline Report: Recycling of Selected Electronic Products in the United States*. Washington, DC: National Safety Council, 1999.

Nelson, Ted. *Computer Lib; Dream Machines*. Redmond, WA: Tempus Books of Microsoft Press, 1987. Originally published as *Computer Lib; Dream Machines: New Freedoms through Computer Screens—a Minority Report* (Chicago: Nelson TH, 1974).

Neville, Brian, and Johanne Villeneuve. "Introduction: In Lieu of Waste." In *Waste-Site Stories: The Recycling of Memory*, edited by. Brian Neville and Johanne Villeneuve, 1–25. Albany: State University of New York Press, 2002.

O'Brien, Martin. *A Crisis of Waste? Understanding the Rubbish Society*. London: Routledge, 2007.

O'Brien, Martin. "Rubbish-Power: Towards a Sociology of the Rubbish Society." In *Consuming Cultures*, edited by Jeff Hearn and Sasha Roseneil, 262–77. Houndsmill, UK: Macmillan Press, 1999.

Odlyzko, Andrew. "The History of Communications and Its Implications for the Internet." 2000. http://www.dtc.umn.edu/~odlyzko/doc/history.communications0.pdf.

Office of Public Sector Information. "Environmental Protection: The Waste Electrical and Electronic Equipment Regulations." U.K. Statutory Instrument No. 3289. December 11, 2006. http://www.opsi.gov.uk/si/si2006/20063289.htm.

Oppenheim, Dennis. Interview, March 29, 1969. In *Recording Conceptual Art*, edited by Alexander Alberro and Patricia Norvell, 21–30. Berkeley: University of California Press, 2001.

Osborne, Peter. "Small-Scale Victories, Large-Scale Defeats: Walter Benjamin's Politics of Time." In *Walter Benjamin's Philosophy: Destruction and Experience*, edited by Andrew Benjamin and Peter Osborne, 57–107. Manchester: Clinamen Press, 2000.

Oudshoorn, Nelly, and Trevor Pinch, eds. *How Users Matter: The Co-Construction of Users and Technologies*. Cambridge, MA: MIT Press, 2003.

Packard, Vance. *The Waste Makers*. New York: David McKay, 1960.

Parks, Lisa. "Falling Apart: Electronics Salvaging and the Global Media Economy" In *Residual Media*, edited by Charles Acland, 32–47. Minneapolis: University of Minnesota Press, 2007.

Parks, Lisa. "Kinetic Screens: Epistemologies of Movement at the Interface." In *MediaSpace: Place, Scale and Culture in a Media Age*, edited by Nick Couldry and Anna McCarthy, 37–57. London: Routledge, 2004.

Pask, Gordon, and Susan Curran. *Micro Man: Living and Growing with Computers*. London: Century, 1982.

Patterson, Dave. "A Conversation with Jim Gray." *ACM Queue* 1, no. 4 (June 2003): 8–17.

Paulos, Eric, and Tom Jenkins. "Urban Probes: Encountering Our Emerging Urban Atmospheres." *Proceedings of the SIGCHI Conference on Human Factors in Computing Systems*, April 2–7, 2005. Portland, Oregon.

Pellow, David Naguib. "Electronic Waste: The 'Clean Industry' Exports Its Trash." In *Resisting Global Toxics: Transnational Movements for Environmental Justice*, 184–24. Cambridge, MA: MIT Press, 2007.

Pellow, David Naguib, and Lisa Sun-Hee Park. *The Silicon Valley of Dreams: Environmental Injustice, Immigrant Workers, and the High-Tech Global Economy*. New York: New York University Press, 2002.

Peters, John Durham. *Speaking into the Air: A History of the Idea of Communication*. Chicago: University of Chicago Press, 1999.

Pfeiffer, K. Ludwig. "The Materiality of Communication." In *Materialities of Communication*, edited by Hans Ulrich Gumbrecht and K. Ludwig Pfeiffer, translated by William Whobrey, 1–14. Stanford: Stanford University Press, 1994.

Pierce, John Robinson. "The Origins of Information Theory." In *An Introduction to Information Theory: Symbols, Signals and Noise*, 19–44. New York: Dover, 1980.

Powers, Richard. *Gain*. New York: Picador, 1999.

Preda, Alex. "Socio-Technical Agency in Financial Markets: The Case of the Stock Ticker." *Social Studies of Science* 36, no. 5 (October 2006): 753–82.

Pryke, Michael, and John Allen. "Monetized Time-Space: Derivatives—Money's 'New Imaginary?'" *Economy and Society* 29, no. 2 (May 2000): 264–84.

Puckett, Jim. "High-Tech's Dirty Little Secret: The Economics and Ethics of the Electronic Waste Trade." In *Challenging the Chip: Labor Rights and Environmental Justice in the Global Electronics Industry*, edited by Ted Smith, David A. Sonnenfeld, and David Naguib Pellow, 225–33. Philadelphia: Temple University Press, 2006.

Raley, Rita. "eEmpires." *Cultural Critique* 57 (Spring 2004): 111–50.

Rabinow, Paul. *Marking Time*. Princeton: Princeton University Press, 2007.

Rathje, William, and Culleen Murphy. *Rubbish! The Archaeology of Garbage*. New York: HarperCollins, 1992.

Reyes, Paul. "Bleak Houses: Digging through the Ruins of the Mortgage Crisis." *Harper's*, October 2008, 31–45.

Rogers, Heather. *Gone Tomorrow: The Hidden Life of Garbage*. New York: New Press, 2005.

Rosenberg, Daniel, and Susan Harding, eds. *Histories of the Future*. Durham: Duke University Press, 2005.

Rudwick, Martin. *The Meaning of Fossils: Episodes in the History of Paleontology*. Chicago: University of Chicago Press, 1976.

Rugemer, Werner. "The Social, Human, and Structural Costs of High Technology: The Case of Silicon Valley." *Nature, Society, and Thought*, 1 (1987): 149–60.

Saar, Steven, and Valerie Thomas. "Toward Trash That Thinks: Product Tags for Environmental Management." *Journal of Industrial Ecology* 6, no. 2 (2003): 133–46.

Sassen, Saskia. "Embeddedness of Electronic Markets." In *The Sociology of Financial Markets*, edited by Karin Knorr-Cetina and Alex Preda, 17–37. Oxford: Oxford University Press, 2005.

Scanlan, John. *On Garbage*. London: Reaktion Books, 2005.

Schiller, Dan. *Digital Capitalism: Networking the Global Market System*. Cambridge, MA: MIT Press, 1999.

Schröter, Jens. "Archive—Post/photographic." Media Art Net. http://www.medienkunstnetz.de/themes/photo_byte/archive_post_photographic/.

Sebald, W. G. *The Rings of Saturn*. Translated by Michael Hulse. 1995. London: Vintage, 1998.

Seitz, Frederick, and Norman G. Einspruch. *Electronic Genie: The Tangled History of Silicon*. Urbana: University of Illinois Press, 1998.

Sekula, Allan. *Fish Story*. Düsseldorf: Richter Verlag, 1995.

Sellen, Abigail, and Richard Harper. *The Myth of the Paperless Office*. Cambridge, MA: MIT Press, 2002.

Semiconductor Industry Association. http://www.sia-online.org/cs/industry_resources/industry_fact_sheet.

Serres, Michel. *Genesis*. Translated by Genevieve James and James Nielson. Ann Arbor: University of Michigan Press, 1995.

Serres, Michel. *Hermes: Literature, Science, Philosophy*. Edited by Josue V. Harari and David F. Bell. Baltimore: Johns Hopkins University Press, 1982.

Serres, Michel. *The Natural Contract*. Translated by Elizabeth MacArthur and William Paulson. Ann Arbor: University of Michigan Press, 1995.

Serres, Michel. *The Parasite*. Translated by Lawrence R. Scher. Baltimore: Johns Hopkins University Press, 1982.

Serres, Michel. *Rome*. Translated by Felicia McCarren. Stanford: Stanford University Press, 1991.

Serres, Michel, with Bruno Latour. *Conversations on Science, Culture, and Time*. Translated by Roxanne Lapidus. Ann Arbor: University of Michigan Press, 1995.

Shanks, Michael. *Experiencing the Past: On the Character of Archaeology*. London: Routledge, 1992.

Shanks, Michael, David Platt, and William L. Rathje. "The Perfume of Garbage: Modernity and the Archaeological." *Modernism/Modernity* 11, no. 1 (2004): 61–83.

Shannon, Claude E. "The Mathematical Theory of Communication." In *The Mathematical Theory of Communication*, by Claude E. Shannon and Warren Weaver, 29–125. 1949. Reprint, Urbana: University of Illinois Press, 1963.

Shannon, Claude E., and Warren Weaver. *The Mathematical Theory of Communication*. 1949. Reprint, Urbana: University of Illinois Press, 1963.

Shiller, Robert J. *Irrational Exuberance*. Princeton: Princeton University Press, 2000.
Shred Tech. http://www.shred-tech.com/electronic.html.
Silicon Valley Toxics Coalition. "Electronics Lifecycle." http://www.svtc.org/site/PageServer?pagename=svtc_lifecycle_analysis.
Silicon Valley Toxics Coalition. "Green Chemistry." http://www.svtc.org/site/PageServer?pagename=svtc_green_chemistry.
Silicon Valley Toxics Coalition. http://www.svtc.org.
Silicon Valley Toxics Coalition. "Silicon Valley Toxic Tour." http://www.svtc.org/site/PageServer?pagename=svtc_silicon_valley_toxic_tour.
Simmel, Georg. "The Metropolis and Mental Life." In *Simmel on Culture*, edited by David Frisby and Mike Featherstone, 174–86. 1903. Reprint, London: Sage, 1997.
Simmel, Georg. *The Philosophy of Money.* Translated by Tom Bottomore and David Frisby. 1907. Reprint, London: Routledge, 1990.
Slade, Giles. *Made to Break: Technology and Obsolescence in America*. Cambridge: Harvard University Press, 2006.
Slater, Don. "Markets, Materiality and the 'New Economy.'" In *Market Relations and the Competitive Process*, edited by Stan Metcalfe and Alan Warde, 95–113. Manchester: Manchester University Press, 2002.
Slater, Don, and Andrew Barry. Introduction to *The Technological Economy*, edited by Don Slater and Andrew Barry, 1–27. London: Routledge, 2005.
Smith, Jeffrey W., James P. Selway III, and D. Timothy McCormick. "The Nasdaq Stock Market: Historical Background and Current Operation." NASD Working Paper 98-01, NASD Economic Research Department: Washington, DC, 1998.
Smith, Ted, David A. Sonnenfeld, and David Naguib Pellow, eds. *Challenging the Chip: Labor Rights and Environmental Justice in the Global Electronics Industry*. Philadelphia: Temple University Press, 2006.
Solving the E-waste Problem (StEP). "Annual Report." 2009. http://www.step-initiative.org/pdf/annual-report/Annual_Report_2009.pdf.
Star, Susan Leigh, ed. *The Cultures of Computing*. Oxford: Blackwell, 1995.
Steedman, Carolyn. *Dust: The Archive and Cultural History*. Manchester: Manchester University Press, 2001.
Sterling, Bruce. "Built on Digital Sand: A Media Archaeologist Digs the Lonely Shores of Binary Obsolescence." In "Ghost: Archive, Evolution, Entropy," special issue, *Horizon Zero*, issue 18 (2004). http://www.horizonzero.ca/textsite/ghost.php?is=18&file=4&tlang=0.
Sterling, Bruce. "The Dead Media Project." http://www.deadmedia.org. Accessed March 4, 2008.
Sterling, Bruce. "The Dead Media Project: A Modest Proposal and a Public Appeal." http://www.deadmedia.org/modest-proposal.html. Accessed March 4, 2008.
Sterling, Bruce. *Shaping Things*. Cambridge, MA: MIT Press, 2005.
Sterne, Jonathan. "Out with the Trash: On the Future of New Media." In *Residual Media*, edited by Charles Acland, 16–31. Minneapolis: University of Minnesota Press, 2007.
Stewart, Kathleen. *A Space on the Side of the Road.* Princeton: Princeton University Press, 1996.
Stiegler, Bernard. *Technics and Time, 1: The Fault of Epimetheus.* Translated by Richard Beardsworth and George Collins. Stanford: Stanford University Press, 1998.
Strasser, Susan. *Waste and Want: A Social History of Trash*. New York: Metropolitan Books, 1999.

Strathern, Marilyn. *Partial Connections.* Savage, MD: Rowman and Littlefield, 1991.
Strathern, Marilyn. *Property, Substance and Effect: Anthropological Essays on Persons and Things.* London: Athlone Press, 1999.
Straw, Will. "Exhausted Commodities: The Material Culture of Music." *Canadian Journal of Communication* 25, no. 1 (2000). http://www.cjc-online.ca/index.php/journal/article/viewArticle/1148/1067.
Suchmann, Lucy. *Plans and Situated Actions: The Problem of Human-Machine Communication.* Cambridge: Cambridge University Press, 1987.
Swiss Agency for the Environment, Forests and Landscape. "Ordinance on the Return, the Taking Back and the Disposal of Electrical and Electronic Equipment (ORDEE)." January 14, 1998. http://www.bafu.admin.ch/abfall/01472/01482/index.html?lang=en.
Taussig, Michael. *Mimesis and Alterity: A Particular History of the Senses.* London: Routledge, 1993.
Taussig, Michael. *My Cocaine Museum.* Chicago: University of Chicago Press, 2004.
Thomas, Valerie. "Radio-Frequency Identification: Environmental Applications." White Paper, Foresight in Governance Project. Washington, DC: Woodrow Wilson International Center for Scholars, 2008.
Thompson, Michael. *Rubbish Theory: The Creation and Destruction of Value.* Oxford: Oxford University Press, 1979.
Thrift, Nigel. *Knowing Capitalism.* London: Sage, 2005.
Thrift, Nigel. "Performing Cultures in the New Economy." *Annals of the Association of American Geographers* 90, no. 4 (2000): 674–92.
Thrift, Nigel, and Shaun French. "The Automatic Production of Space." *Transactions of the Institute of British Geographers* 27, no. 3 (2002): 309–35.
Toffler, Alvin. *Future Shock.* New York: Random House, 1970.
Tomlinson, Bill. *Greening through IT: Information Technology for Environmental Sustainability.* Cambridge, MA: MIT Press, 2010.
Toxics Link. "Fact File on Waste in a Wireless World." Delhi: Toxics Link, February 2003.
Toxics Link. http://www.toxicslink.org.
Toxics Link. "Scrapping the Hi-Tech Myth—Computer Waste in India." Delhi: Toxics Link, February 2003.
Trash Track. http://senseable.mit.edu/trashtrack.
Turing, Alan. "Intelligent Machinery." In *Mechanical Intelligence,* edited by D. C. Ince, 107–28. 1948. Repr., London: North Holland, 1992.
"UN Programme Aims at Environmentally Sound Disposal of Electronic Waste." UN News Centre, November 25, 2005. http://www.un.org/apps/news/story.asp?NewsID=16690&Cr=electronic&Cr1=.
United Nations Environment Programme. "Basel Conference Addresses Electronic Wastes Challenge." November 27, 2006. http://www.unep.org/Documents.Multilingual/Default.asp?DocumentID=485&ArticleID=5431&l=en.
United Nations Environment Programme. "Guidelines for Social Life Cycle Assessment of Products." DTI/1164/PA. http://www.unep.org/pdf/DTIE_PDFS/DTIx1164xPA-guidelines_sLCA.pdf. 2009.
United Nations Environment Programme. "Recycling: From E-Waste to Resources." United Nations Environment Programme and United Nations University, 2009. DTI/1192/PA. http://www.unep.org/PDF/PressReleases/E-Waste_publication_screen_FINALVERSION-sml.pdf.

United Nations Environment Programme. "Vital Waste Graphics." UNEP, 2004.
United Nations Environment Programme. "Vital Waste Graphics, 2." 2nd ed. UNEP, 2006.
United Nations University. "Review of Directive 2002/96 on Waste Electrical and Electronic Equipment (WEEE)." Tokyo: UNU, 2008.
United States Government Accountability Office. "Electronic Waste: EPA Needs to Better Control Harmful U.S. Exports through Stronger Enforcement and More Comprehensive Regulation." GAO-08-1044. August 2008.
Variable Media Network. http://www.variablemedia.net.
Voet, Ester van der, Lauran van Oers, and Igor Nikolic. "Dematerialization: Not Just a Matter of Weight." *Journal of Industrial Ecology* 8, no. 4 (2005): 121–37.
Watkins, Evan. *Throwaways: Work Culture and Consumer Education*. Stanford: Stanford University Press, 1993.
Weaver, Warren. "Some Recent Contributions to the Mathematical Theory of Communication." In *The Mathematical Theory of Communication*, edited by Claude E. Shannon and Warren Weaver, 1–28. 1949. Reprint, Urbana: University of Illinois Press, 1963.
Wiener, Norbert. *Cybernetics: Or Control and Communication in the Animal and the Machine*. Cambridge, MA: MIT Press, 1961.
Wiener, Norbert. *The Human Use of Human Beings*. New York: Doubleday Books, 1954.
Willis, Anne-Marie. Editorial: De/re/materialization (contra-futures). *Design Philosophy Papers* 2 (2005). http://www.desphilosophy.com.
Woodward, Kathleen, ed. *The Myths of Information: Technology and Postindustrial Culture*. Madison, WI: Coda Press, 1980.
Zaloom, Caitlin. "Ambiguous Numbers: Technology and Trading in Global Financial Markets." *American Ethnologist* 30, no. 2 (2003): 258–72.
Zaloom, Caitlin. *Out of the Pits: Traders and Technology from Chicago to London*. Chicago: University of Chicago Press, 2006.
Zehle, Soenke. "Environmentalism for the Net 2.0." *Mute: Culture and Politics after the Net*. 21 (September 2006). http://www.metamute.org/en/Environmentalism-for-Net-2.0.
Zielinski, Siegfried. *Deep Time of the Media: Toward an Archaeology of Hearing and Seeing by Technical Means*. Translated by Gloria Custance. Cambridge, MA: MIT Press, 2006.

Newspaper Articles

"Abandoned in the E-waste Land." *Observer*, February 2, 2003. http://www.guardian.co.uk/society/2003/feb/02/business.conferences3.
Adams, Jane Meredith. "Silicon Valley's Tech Waste Problem." *Chicago Tribune*, January 28, 2003.
Borin, Elliott. "Junked PCs Offer Data for Taking." *Wired News*, September 25, 2002.
Branigan, Tania. "From East to West, a Chain Collapses." *Guardian*, January 9, 2009.
"Cardboard PC Case by Lupo." UberGizmo, October 21, 2005. http://www.ubergizmo.com/15/archives/2005/10/cardboard_pc_ca.html.
Cool Light Leads to Greener Chips." *BBC News*, June 30, 2006. http://news.bbc.co.uk/1/hi/technology/5128762.stm.
Dean, Katie. "Disposable DVDs at Crossroads." *Wired News*, February 7, 2005. http://www.wired.com/entertainment/music/news/2005/02/66513.

Fildes, Jonathan. "The Winds of Change for Africa." *BBC News*, July 23, 2009.
Fildes, Jonathan. "Wireless Power System Shown Off." *BBC News*, July 23, 2009.
Fisher, Jim. "Poison Valley: Is Workers' Health the Price We Pay for High-Tech Progress?" *Salon*, 2001. http://archive.salon.com/tech/feature/2001/07/30/almaden1/.
Glaister, Dan. "US Recycling: 'I Don't Even Think We Have an Industry.'" *Guardian*, January 9, 2009.
"Greenpeace Deploys GPS to Track Illegal Electronic Waste." Environment Blog. *Guardian*, February 18, 2009. http://www.guardian.co.uk/environment/blog/2009/feb/18/greenpeace-electronic-waste-nigeria-tv-gps.
Greenpeace News. "Undercover Operation Exposes Illegal Dumping of E-waste in Nigeria." February 18, 2009. http://www.greenpeace.org/international/en/news/features/e-waste-nigeria180209.
Hickman, Leo. "The Truth about Recycling." *Guardian*, February 26, 2009.
"Intel's Antitrust Ruling: A Billion-Euro Question." *Economist*, May 14, 2009.
Johnson, Bobbie. "Google's Power-Hungry Data Centres." *Guardian*, May 3, 2009.
Johnson, Bobbie. "Web Providers Must Limit Internet's Carbon Footprint, Say Experts." *Guardian*, May 3, 2009.
Joy, Bill. "Why the Future Doesn't Need Us." *Wired News*, issue 8, no. 4. April 2000.
Leahy, Stephen. "Short-Lived PCs Have Hidden Costs." *Wired News*, March 8, 2004.
Oliver, Christine. "Recycling in the Credit Crunch." *Guardian*, January 9, 2009.
Onion, Amanda. "Buy, Use, Dispose: A Spike in Disposable Products Has Environmentalists Worried." *ABC News*, December 4, 2002.
Osborne, Hilary. "Rich Nations Accused of Dumping E-Waste on Africa." *Guardian*, November 27, 2006.
Pearce, Fred. "Greenwash: WEEE Directive Is a Dreadful Missed Opportunity to Clean up E-waste." *Guardian*, June 25, 2009.
Ritchel, Matt. "E.P.A. Takes Second Look at Many Superfund Sites." *New York Times*, January 31, 2003.
Rivlin, Gary. "In Silicon Valley, Millionaires Who Don't Feel Rich." *New York Times*, August 5, 2007.
Schofield, Jack. "When the Chips are Down." *Guardian*, July 29, 2009.
Shabi, Rachel. "The E-waste Land." *Guardian*, November 30, 2002.
"Spammers Manipulate Stock Markets." *BBC News*, August 25, 2006. http://news.bbc.co.uk/go/pr/fr/-/2/hi/technology/5284618.stm.
Wray, Richard. "Spam 'Uses as Much Power as 2.1M Homes.'" *Guardian*, April 15, 2009.

Archives and Museums

Charles Babbage Institute (The Diebold Group collection, 1957–90), University of Minnesota.
Computer History Museum (and Archive), Mountain View, CA.
Computing History Archive (ICL collection), Science Museum of London.
Computing History Archive, Smithsonian Institute, Washington, DC.
Intel Museum, Santa Clara, CA.
National Archive for the History of Computing, Manchester, UK.
Technology Museum, San Jose, CA.

Index

Access Space, 154
Acland, Charles, 9
actor-network theory, 164n45, 189n44
Adkins, Lisa, 174n4
Adobe, 39
Agarwal, Ravi, and Kishore Wankhade, 134
Akrich, Madeline, 163n29
Alexander, Judd H., 196n10
Alexandrian Library, 107
Amdahl, Gene, 101, 104
Appadurai, Arjun, 190n61
appliances, 80–81, 83, 86
Architectural Record, 52
archives, electronic, 13, 15, 36–37, 98, 105, 107–13, 118–25, 130, 154, 186n1
automation, 12, 14, 47, 49, 60, 62, 63–65, 69, 80–83, 93, 182n18, 189n43

Babbage, Charles, 63, 114
Barney, Darin, 171n50, 178n52
Barthes, Roland, 86
Basel Action Network, 69, 94, 129, 139, 141–42, 184n49, 193n17, 194nn38–39
Baudelaire, Charles, 192n7
Baudrillard, Jean, 41, 181n15
Bauman, Zygmunt, 95, 183n39, 185n68
Bell, Daniel, 169n21
Beniger, James, 33, 170nn38–39
Benjamin, Walter, 106, 131–32, 151, 155–56, 161n17, 161n19, 162n21, 162n26, 163n31, 187n14, 192n82, 192n7, 193n24, 196n56; *The Arcades Project,* 1, 5–9, 11–13, 17, 104–5, 161n19, 165n57, 174n11, 176n35, 182n16, 186nn3–4
Bennett, Jane, 200n40
Bensaude-Vincent, Bernadette, and Isabelle Stengers, 85
black box, vi, 62, 70, 154, 159n1
Bolter, J. David, 74, 187n10
Boyle, James, 199n36
Brown, Bill, 161n13, 192n6
bubbles, financial, 49, 55, 68, 174n5, 175n22
Buchli, Victor, 121, 142, 155, 166n68, 191n72
bullets, 104–5
Bush, Vannevar, 36–37, 108–9
Butler, Judith, 8, 9, 161n12, 161n17

calculation, 33–36, 38, 49–50, 171nn50–51, 172n57, 175n22
Callon, Michel, 176n25, 190n57
Calvino, Italo, 74, 77–78, 89, 97–98, 180–81n1
cathode-ray tubes, 70, 91, 137, 140, 180n88
Ceruzzi, Paul, 113, 116, 189nn51–53
chemical industry, 25–27, 39
Chun, Wendy Hui Kyong, 188n38
circulation, 13, 66–69, 77–78, 86, 136, 139, 166n60, 179n70
Clark, Gordon, and Nigel Thrift, 56
Clarke, Arthur C., 108
CNBC, 55
CNN, 55
Colloredo-Mansfeld, Rudi, 79, 97
commodities, 8, 9, 11, 12, 47, 66, 68, 78–79, 93, 114, 136, 190n61. *See also* things
compression, 36–37, 108, 172n53
Computer History Museum, 101
computers, 20, 33, 51, 64, 70, 87, 101, 108, 110, 113–17, 124, 133, 171n50, 177n44, 189n53
conservation, 121
consumption, 4–5, 12, 16, 17, 28, 33, 59, 68, 77, 78, 81–83, 96–98, 129, 155, 182n17, 185n65

copper, 3–4, 13, 70, 75, 87, 91, 131, 134–35, 137, 180n88
Cornucopia City. *See* Packard, Vance
Corzo, Miguel Angel, 110, 187n22

Darwin, Charles, 6, 162n24
dead/failed media, v–vii, 2, 5, 6–7, 15, 98, 101–25
Deleuze, Gilles, 166n60
dematerialization, 4, 7, 14, 42, 47, 48, 50, 51, 53–54, 57–61, 69–70, 86, 88, 132, 139, 142, 172n58, 175n15, 180n84, 191n74
Dick, Philip K., 127
Diebold, John, 63–64, 118, 182n18
digital media, vii, 2, 10, 16, 36, 86, 110–11, 140
digitization, 24, 32, 87, 110, 121, 170n38, 178n59
dirt, 17, 89, 96, 143, 155, 183n40, 185n72
discourse, defined, 161n12
disposability, 74, 77–98, 125
dot-com crash, 48–49, 55, 68, 167n5
Douglas, Mary, 183n43
dumps, 127–46; in Africa, 15, 61, 91, 94, 129, 161n11; in China, x, 15, 61, 75, 90, 91, 94, 98, 128, 144–45, 185n61; in India, 15, 90, 94, 134, 185n61; in southeast Asia, 69–70, 91. *See also* landfills
Dunne, Anthony, 199n30
Dupuy, Jean-Pierre, 180n86
Dyer-Witheford, Nick, 175n14

ecologies, 2, 11, 16, 70, 137, 154, 159n3, 199n35
electronic trading, 48–58, 61–64, 69, 175n12, 175–76nn22–23, 178n55
emulation systems, 111, 123–24, 188n26
Engler, Mira, 141
ENIAC, 63
entropy, 33, 171n42
environmentalism, 154, 199n36, 200n38
environmental overload, 38–41
Environmental Protection Agency, 1, 39, 194n28
ephemerality. *See* disposability; obsolescence
Ernst, Wolfgang, 119, 124, 190n64, 190n69, 191n74, 192n86
exchange, 45, 47–51, 54–58, 63–71, 95–97, 164n45, 178n55, 179n69, 190n61
expenditure, 54–56, 60, 79
"extended producer responsibility," 91

fabs (fabrication facilities), 22, 27, 39, 168nn14–15
failure, 104, 106, 111, 114, 124–25, 162n25, 190n58. *See also* Museum of E-Failure; obsolescence
Fairchild Semiconductor, 25, 30–31, 39
feral robotic dogs, 199n30
Fisher, Jim, 167n1
Fisher, Melissa, and Greg Downey, 174n5
Fladager, Vernon, 81
Forty, Adrian, 182n27, 191n73
"fossils." *See* "natural history" methodology
Foucault, Michel, 6, 166n60
Franklin, Sarah, Celia Lury, and Jackie Stacey, 163n36
Frow, John, 68, 136
Fry's Electronics Superstore, 19, 29
Fuller, Matt, 159n3
Furtherfield, 154

Gandy, Matthew, 194n28
Gaonkar, Dilip Parameshwar, and Elizabeth A. Povinelli, 166n60
garbage imaginary, 155–56
Garbage Project, 16
garbology, 16, 17, 127, 131, 140. *See also* rubbish theory
General Electric, 63
Gitelman, Lisa, 188n38
gold, 91, 129, 131, 134, 137
Google, 48, 172n53, 177n48
Gould, Stephen Jay, 162n24
green electronics, 151–54, 198n23
Greenpeace, 91, 151, 193n17
Greenspan, Alan, 55
Guattari, Felix, 199n35

Haggerty, Patrick, 31
Hansen, Karen Tranberg, 193n21
Hansen, Mark, 187n14
Hanssen, Beatrice, 162n21
Haraway, Donna, 1, 8, 12, 20, 45, 50, 57, 70, 164nn45–46, 167
hardware, discarded, vi, 2, 3, 5, 13, 38, 70–71, 74–98, 135, 140, 154
Harvey, David, 173n63, 174n7

Harwood, Graham, and Matsuko Yokokoji, 154
Hawkins, Gay, 82, 150–51, 194n28
Hayles, N. Katherine, 123, 161n16, 170n36, 172n58, 177n41
hazardous waste, vii, 1, 3, 4, 11, 15, 27, 70, 95, 129, 136, 139, 142, 151, 160n9, 183n37, 184n54, 195n47, 197n13. *See also* pollution
Hetherington, Kevin, 89, 182n17, 185n72, 186n7
Hinte, Ed van, 152
"How Much Information," 34–36, 66

IBM, 25, 104
ICL, 103, 105, 126, 157
information overload, 4, 11, 14, 32–38
information theory, 10, 29, 32–33, 66, 170n36, 171n42
"informed materials," 85
Intel, 25, 26, 30–31, 39, 48, 84, 117, 169n31, 173n65, 181n10
International Association of Electronics Recyclers, 133, 160n5, 160n8
Internet, 35, 61, 67, 114, 122–23, 154, 177n48
Internet Archive, 122–23

Jeremijenko, Natalie, 199n30

Keenan, Thomas, 188n38
Kelly, Kevin, 59–61, 65, 67
Kilby, Jack, 169n24
Kittler, Friedrich, 10, 62, 110, 124, 178n59, 187n19, 188n26
Knorr-Cetina, Karin, and Urs Bruegger, 177n38
Koolhaas, Rem, 143
Krauss, Rosalind, 186n4
Krishna, Gopal, 2

landfills, 2, 13, 15–16, 88, 89, 90–91, 127, 130–31, 140–42, 151, 154, 194–95n39
Lécuyer, Christophe, and David C. Brock, 167n3
Leonia. *See* Calvino, Italo
Leslie, Esther, 28, 168n18
London, Bernard, 189n42
Long Now Foundation, 140, 189n43
Loon, Joost Van, and Ida Sabelis, 138, 150–51, 193n9, 195n50, 196nn5–6, 197nn11–12
Lucas, Gavin, 97
Lynch, Kevin, 141, 147, 155
Lyotard, Jean François, 33–34, 185n72

Machlup, Fritz, 66–67
MacKenzie, Donald, 116–17, 190n57–59
Manzini, Ezio, 84, 87, 182n21
Marx, Karl, 41, 96–97, 173n65, 174n7, 174–75n11, 176n35
material flows, 26, 79, 94, 175n15, 180n85, 181n6
materiality, vi, 3–4, 7, 9, 10, 14, 17, 47, 56, 58–59, 79, 88, 117, 131, 132, 139, 150, 151, 153, 165n51, 167n3, 177nn40–41
McLuhan, Marshall, 10, 65, 159n3
media archaeology, 164n41
media studies, 9–10, 17, 171n50
Memex, 37, 108–9
memory, electronic, 101, 105, 107–13, 119, 121, 125, 187n8
metals, vi, 3, 28, 87, 91, 121, 137
microchips, v, vi, 1, 3, 5, 7–8, 12, 20–32, 38–39, 41, 48, 51, 53, 58, 62, 78, 80–81, 86–87, 107, 117, 134–35, 152, 168n12, 169n25
Microsoft, 48, 109
mining, vi, 165n59, 196n10
Miyazaki, Hirokazu, 165n51
mobility, 88–89, 93, 95
Modern Packaging Journal, 85
Molotch, Harvey, 111, 151, 181n10, 194n28
monitors. *See* screens
Moore, Gordon, 30–31, 117, 169nn29–30
Moore's Law, 27, 30, 116–17, 121, 152, 169n29, 173n65
Moser, Walter, 17, 185n70, 196n7
Mulder, Arjen, 188n26
Museum of E-Failure, 122
museums, 101, 104, 107, 116, 122, 125, 186n1, 186n7
MyLifeBits, 109–10, 120, 188n24, 190n69

NASDAQ, 13, 14, 46–56, 59, 61–63, 68, 154, 173n2, 177n48
NASDAQ MarketSite, 46, 52–55, 70
"natural history" methodology, vii, 5–13, 17, 65, 106, 118, 149–50, 153, 161n19, 162n24, 162–63nn27–28

nature, 1, 7, 8–9, 149, 150, 155, 162n21, 163n28, 163n34, 163n36, 194n28, 200n40
Nelson, Ted, 101
Netscape, 39
networks, electronic, 59–62, 64–67, 112, 164n45, 178nn52–53
Neville, Brian, and Johanne Villeneuve, 192n3
"new economy," defined, 174nn4–5
"new media," 114, 188n38
New York Stock Exchange, 45, 48, 52, 63–64
Noyce, Robert, 30–31, 169n24

O'Brien, Martin, 185n71, 195n44
obsolescence, vii, 2, 4, 7, 10, 15, 17, 78, 82–84, 97, 113–18, 121, 149, 152, 182n15, 186n4, 188n39, 189n42, 189nn44–45
offshoring, 29, 40, 168n14, 169n23, 180n86, 185n61
open source software, 154
Oppenheim, Dennis, 45
Osborne, Peter, 196n56

packaging, 77, 80–82, 84–88, 93, 182n24
Packard, Vance, 77, 82, 92, 97–98, 114, 188n39
Parks, Lisa, 70
Pellow, David Naguib, and Lisa Sun-Hee Park, 168n10
performativity, 4, 47–51, 54, 58–59, 63, 69, 84, 161n12, 174n9, 174n11, 175n22, 176n23, 176n35, 177n40
Peters, John Durham, 170n36
Pfeiffer, K. Ludwig, 177n40
Phillips, Bill, 179n74
Pierce, John Robinson, 171n42
plastics, 3, 13, 14, 15, 24, 28, 77, 80–81, 84–88, 93, 94, 121, 151
pollution, v, 1–2, 3, 15, 20, 24, 25, 28, 32, 39, 40, 87, 95, 96, 141, 142
postmodernism, 171n44, 175n11
Proboscis, 199n30
progress, concepts of, 6–7, 9, 12, 104–6, 121, 162n26, 196n56, 197n11
Pryke, Michael, and John Allen, 175n12
Puckett, Jim, 195n47

Rabinow, Paul, 188n38
Raley, Rita, 50, 178n53
Rathje, William, 16, 155, 182n24
Raytheon, 25
recycling, viii, 2–3, 15, 16, 41, 61, 69–70, 77, 87, 89–97, 100, 129, 132–42, 150–51, 183n36, 184nn49–50, 185n70, 194n28, 194–95nn38–39
rematerialization, 70, 87, 132, 165n52
repair, 82, 133, 193n12
resonance, defined, 164n46
Reyes, Paul, 180n80
Rogers, Heather, 92–93
RoHS, 160n8, 184n52, 198n22
RSA WEEE Man, 76, 80
rubbish theory, 16–17, 49, 68, 133. *See also* garbology
Rudwick, Martin, 164n36
Rugemer, Werner, 168n17

salvaging, 70, 87, 95, 129–37, 139–40, 143, 183n36, 192–93nn7–8, 193n21
Sans Soleil, 188n25
Scanlan, John, 183n39, 195n52
scatter, 8, 151, 163n31
Schiller, Don, 60, 178n51
Schröter, Jens, 123, 191n78–79
screens, 5, 13, 14, 42, 45–48, 52–59, 62, 68, 70, 73, 78, 177n38
sedimentation, defined, 161n17
Sekula, Alan, 93–94
Serres, Michel, 17, 45, 47, 65, 84, 147, 179n69, 185n72, 189n45
Shanks, Michael, 131; with David Platt and William L. Rathje, 155
Shannon, Claude, and Warren Weaver, 32–33, 167n2
Shiller, Robert, 55, 176n34
shipping, 90–94, 129, 135, 183n47, 184n54, 185n61, 185n64
Siemens, 25
silicon, 3, 20–21, 24, 26–28, 38, 41–42, 84–85
Silicon Fen, 40
Silicon Mountain, 40
Silicon Valley, xii, 1–2, 13, 16, 18–20, 25, 27–28, 29, 38–41, 44, 49, 61, 154, 160n9, 167n6, 172–73n62, 173n65
Simmel, Georg, 172n51, 172n59
Slade, Giles, 183n47, 189n42
Slater, Don, 58; and Andrew Barry, 172n51
"social death," 79

software, v–vi, 3, 62–63, 154
solvents, 3, 25
spam, 67–68, 179nn78–79
speed, 25–31, 48–50, 55–58, 64, 69, 81, 105, 114, 176n35, 178n55, 187n10, 189n43
St. Mungo's, 154
Steedman, Carolyn, 139
Sterling, Bruce, v, 5, 111–12, 113, 115, 199n28
Sterne, Jonathan, 188n38
Stevens, Brooke, 189n42
Stewart, Kathleen, 166n58
Stiegler, Bernard, 101, 175n22, 188n33
storage, electronic. *See* archives, electronic; memory, electronic
Strasser, Susan, 183n43
Strathern, Marilyn, 142
Straw, Will, 104, 166n64, 189n48
Superfund sites, 1, 13, 16, 18, 20, 25, 27, 39, 41, 61, 159n1, 167n6, 172nn61–62
sustainable computing, 151–54, 198nn23–26, 199nn32–33

Taussig, Michael, 136–37
temporality, 6–7, 10, 49–51, 57, 81, 93, 101–25, 130–31, 138–39, 155–56, 195n50, 197n11
Texas Instruments, 31, 84
things, 8, 47–49, 58–59, 68, 78–79, 83, 96–98, 106, 111, 127, 131, 136, 139, 142–43, 147–49, 153, 155, 165n58, 166n60, 177n40, 179n70, 185n72, 192n6. *See also* commodities
Thompson, Michael, 16–17, 68, 133, 174n6
Thrift, Nigel, 56, 176n25
"throwaway society," 80, 82–84
Toffler, Alvin, 88, 114–15, 118, 190n62
Tomlinson, Bill, 198n23
Toxics Link, 134, 193n17
toxic waste. *See* hazardous waste; pollution
Turing machine, 63
2007–8 financial crash, 50, 68, 194n29

users, 10, 62, 70, 78, 109, 165n47; using up, 74, 78–79, 96–98, 165n47

value, 2, 16–17, 47–51, 56–59, 65–69, 89–91, 95–96, 133–36, 174n6, 179n70, 185n70, 195n44
Variable Media Network, 191n80
virtuality, vi, 4, 54, 57, 61, 68, 69, 86, 175n20

waste, 2–4, 14, 16–17, 28–32, 47, 49, 51, 61, 67, 68–70, 77, 79, 82, 89, 96, 107, 119, 124–25, 129, 132, 138, 143, 147–50, 155, 183–84n47, 192n3, 194n28, 196n7, 197n13
waste export, 15, 69–70, 78, 91–94, 129, 134–36, 139, 151, 154, 180n86, 184n54
Watkins, Evan, 115
Weaver, Warren, 171n42
Web 2.0, 114, 152
WEEE, 160n8, 184n52, 198n22
Wham-O, 190n62
Wiener, Norbert, 170n36
Willis, Anne-Marie, 180n84
Wired, 59
WISC, 101–2, 104–5
World Resources Institute, 79, 180n85, 181n6

Yahoo!, 39

Zaloom, Caitlin, 57
Zehle, Soenke, 154, 199–200n36
Zero Dollar Laptop, 154
zero waste, 149–50
Zielinksi, Siegfried, 9–10